Skyscrapers and the men who Build Them

SKYSCRAPERS
AND THE MEN WHO BUILD THEM

From a photograph by Pirie MacDonald

SKYSCRAPERS

AND THE MEN WHO BUILD THEM

BY

COL. W. A. STARRETT

MEMBER AMERICAN SOCIETY CIVIL ENGINEERS, MEMBER AMERICAN SOCIETY
MECHANICAL ENGINEERS

AUTHOR OF "MARKED 'SHOP,'" ATLANTIC MONTHLY, JULY, 1917; "BUILDING FOR VICTORY."
SCRIBNER'S, NOVEMBER. 1918; "NEW CONSTRUCTION IN AN ANCIENT EMPIRE,"
SCRIBNER'S, SEPTEMBER, 1923, ETC.

WITH ONE HUNDRED AND THIRTY-TWO ILLUSTRATIONS

CHARLES SCRIBNER'S SONS

NEW YORK ⋅ LONDON

1928

CONTENTS

ILLUSTRATIONS

FOREWORD

In the preparation of this volume the author has made no attempt to present a technical treatise on building, nor has he had any thought of covering all of the aspects of construction work. Many important elements have been referred to only casually, and, in fact, many subjects incidental to skyscraper construction have been omitted altogether.

The purpose of the book is to give the reader a general idea of the planning and construction of a skyscraper, the essentials of its origin, and the administration and functioning of a modern building organization, avoiding wherever possible technicalities that have already been ably covered by architects and engineers in the voluminous bibliography of modern American construction.

The volume has been prepared for the layman who watches with interest, and perhaps fascination, the skyscraper in course of construction and desires to know something of the fundamentals that govern the work without too deeply delving into its technic.

In spite of the vast amount of construction that is everywhere in evidence in our growing cities there seems to be much about the subject that is not generally understood. If the reader's understanding and appreciation of building are quickened by the subject matter of this volume, the author's object will have been attained.

WHISTLES

C. D. CHAMBERLIN

The clean-up gang is finished—the last shanty's coming down;
We've punched our last big payroll out—let's hunt another town.
There's a million smoky whistles, wheezing gods that we obey,
And the order that they're screaming is "Builders! On your way!"
So,
Let's speed—speed—speed!
Out to where the whistles plead,
Wailing at their toiling mob,
Laughing at the lives they rob,
Sneering at the biggest job,
There's work to do. Let's GO!

Jack-hammers singing like the whining choirs of Hell,
Gouging out the bed-rock and a deep foundation well;
Dynamite and steam drills are eating granite rock,
For a city wants a subway and it's work against the clock!
So,
Let's go—go—go!
Out to where the whistles blow,
Yelling out for men to build,
Sobbing for the men they've killed,
Boasting of the jobs they've filled,
There's work to do. Let's GO!

Rivet-guns thundering in an iron-chested roar,
Punching white-hot clinchers on the forty-seventh floor;
Cable-hoists and air lines are stitching boiler plate;
Some one wants a fifty-decker—and they want it while they wait.
So,
Let's climb—climb—climb!

Up to where the whistles chime,
Begging for the men they need,
Roaring at each mighty deed,
There's work to do. Let's GO!

Concrete-mixers rumbling in a heavy, sullen moan,
Chewing rock and gravel up and spitting liquid stone.
Crusher blades and shovels are making boulders bounce;
Quick! A dam to light the cities!—Time is all that counts.
So,
Let's rush—rush—rush!
Where the whistles never hush,
Shrieking shrill their frantic plea,
Shouting loud for you and me,
Booming "Builders! Can't you see
There's *work* to do? Let's GO!

The author of the above poem appeared at Dartmouth College with
his kit of carpenter's tools, and worked his way through by following his
trade; in summer finding employment as a carpenter on large construction work. He wrote the poem while at college in pursuit of his study
of English, and by chance it fell into the hands of the author of this
volume.

The inspiration of the great modern construction job is so completely
caught by the poem that it is here given, with grateful acknowledgment
to the author's genius in producing so vivid a picture of the spirit of
modern building.

SKYSCRAPERS
AND THE MEN WHO BUILD THEM

CHAPTER I

AMERICA CREATES SOMETHING ABSOLUTELY NEW— THE SKYSCRAPER

THE skyscraper is the most distinctively American thing in the world. It is all American and all ours in its conception, all important in our metropolitan life; and it has been conceived, developed and established all within the lifetime of men who are, in many cases, still active in the great calling which they themselves created and which they have developed within the span of their business careers.

I have seen the unfolding of practically the whole drama myself, for while I have just passed the half century in years, I have keen recollections of the building of the first skyscraper that could truly bear the name in the modern acceptance of that term. For the skyscraper, to be a skyscraper, must be constructed on a skeleton frame, now almost universally of steel, but with the signal characteristic of having columns in the outside walls, thus rendering the exterior we see simply a continuous curtain of masonry penetrated by windows; we call it a curtain wall. This seemingly continuous exterior is supported at each floor by the beams or girders of that floor, with the loads carried to the columns embedded in that same masonry curtain, unseen but nevertheless absolutely essential to the towering heights upon which we gaze with such admiration and awe—and pride, our everlasting pride in our completely American creation. We use these skyscrapers and accept them as a matter of course, yet as each new one rears its head, towering among its neighbors, our sense of pride and appreciation is quickened anew, and the metropolis, large or small, wherein it is built, takes it as its very own, and uncomplainingly endures the rattle and roar of its riveting hammers, and the noises and the inconvenience of traffic which it brings. And this is because we recognize it as another of our distinctive triumphs, another token of our solid and material growth.

But our pride of civic acquisition is small compared with the pride we take in our ability to build, for when a great building starts, all the world is a builder and the whole citizenry of a metropolis takes to its heart the swift and skilful accomplishment. The drama as it unfolds excites wonder and admiration, and those of us who have taken part in the creation and production of the drama have a pride and joy that is just what would be imagined by the enthusiastic spectator who gazes with admiration at some feat of skill and daring performed before his very eyes as he looks on from a vantage point, and perhaps sees nature used against its very self in the accomplishment of a spectacular bit of work.

How it all started, and who the men were who brought it all about, is a fascinating tale and one full of dramatic interest. Nations and civilizations may rise and fall and historians of the far distant future may say that we were not many things that we now think we are, but one thing is certain: they will of a surety say that we were a nation of builders, great builders, the greatest that the world had ever seen up to the era of our sudden greatness in construction. Yet a single lifetime can claim to have seen it all from its inception, through its development, and well matured in its accomplishment—distinctive from all other great world accomplishments in that it is, let me say again, essentially and completely American, so far surpassing anything ever before undertaken in its vastness, swiftness, utility, and economy that it epitomizes American life and American civilization, and, indeed, has become the cornerstone and abode of our national progress.

An architectural style, or indeed, many styles, have been created by our great metropolitan skyscrapers, but all are essentially American. The beauties of ancient architecture have been made to serve this new form and the perfections of the ancients have been confirmed by their adaptability to the great solutions of our modern buildings, but the form is ours and it has taken its place with the classics If we never build another building, if we·never make another design, posterity will have to accord us the creation of a great, new architectural style fixed beyond all changing. For America has already been won the title of the greatest builder that the world has ever seen.

For fifty-two years Trinity Church steeple was the zenith of the New York sky-line, two hundred and eighty-four feet above Broadway. Visitors to New York paid their shilling each to climb the three hundred and eight steps, "with suitable resting places provided," to a point thirty-four feet below the peak, and from there looked down on a city limited, as from ancient times, by the feeble power of the unassisted human leg, to a height of six stories; a city hemmed about and bullied by the towering masts of hundreds of sailing ships.

Trinity Church was built in 1841 and was unchallenged until 1893, when, across the street, the Manhattan Life Insurance Building, seventeen stories and tower, thrust its pinnacle sixty feet above Trinity spire. In another fifteen years the steeple had sunk almost without trace beneath the soaring sky-line.

The skyscraper, once possessed of only the beauty of power, has acquired the beauty of line and form as well. Who can look on the majestic sky-line of New York in sunshine or shadow and not be moved, both by the tremendous power of its mass and the beauty and richness of its detail? Yet if you are forty-three years or more old, your eyes have seen it all.

The skyscraper was conceived and demonstrated by a little group of architects, engineers, and builders in one American

city. That city was not New York, with which the world
usually identifies it, but Chicago.

When I sought to confirm and to supplement my own
memory of the history of the skyscraper, I found that no book
ever had been written on its history. So far as I can learn, the
only attempt ever made to trace its origins and development
was done by Robins Fleming, a veteran and scholarly structural
engineer of the American Bridge Company. I am indebted to
Mr. Fleming's researches for many of the facts in this examination of the neglected family tree of an American institution.

The great pioneers of the skyscraper were William Le-
Baron Jenney, Daniel H. Burnham, John W. Root, and William Holabird, pretty much in the order named. Other Chi-
cagoans contributed, but these four were chiefs. Jenney alone
was classically educated in architecture, and he arrived by a
roundabout route. He was born in New Bedford, his father
head of Jenney & Gibbs, a great whale-oil house when the
world still read by the light of the sperm whale's fat. He
left Andover at seventeen and in one of his father's whalers
sailed around the Horn to join the rush of '49 to California.
Later, in the Philippines, he was so struck with the possibilities of railroad building in the islands that he resolved to return home and study engineering.

After three years at Lawrence Scientific School, he went to
Paris and there was an intimate of Whistler and Du Maurier.
In the 50's he was an engineer employed in building the
Panama Railroad. When the Civil War broke out, he joined
the army under General William Tecumseh Sherman and
eventually became Sherman's chief of engineers on the march
from Atlanta to the sea. In 1868 he returned to Chicago and
took up the practice of architecture.

A year or so later, Daniel H. Burnham alighted in that
city from a cattle train from the West. Born at Henderson,
New York, in 1846, he had failed to pass the entrance exami-

Courtesy of Thompson-Starrett Co.

From a photograph by Ewing Galloway.

nations at both Harvard and Yale, and had gone to Nevada on a mad French colonization scheme; he had remained as a miner and a contemporary of Mark Twain in the grand days of Virginia City, had run for state senator, been defeated, and eventually made his way to Chicago as best he could. There he became interested in architecture, sold plate glass in the boom market that followed the great fire in 1871, and finally became a draftsman for Carter, Drake & Wight. Wight was himself a building genius and the works he has left indicate that he visioned the development that was about to follow. The head draftsman there was John W. Root, who was later to become Burnham's partner in the firm of Burnham & Root.

Root was a Georgian, his father a dry-goods merchant, of frustrated longings to be an architect, who had made a fortune in blockade-running during the Civil War. In the midst of the war the son was smuggled out of Wilmington, North Carolina, on a cotton ship that slipped past the guns of the Federal fleet, and was put in school in England. Later he was studying engineering in the College of the City of New York and was headed for the Ecole des Beaux Arts, when the family fortune was lost as quickly as it had been gained and the student went to work in a Chicago architect's office. John W. Root died in 1891 in the first promise of a great career.

William Holabird was the son of General Samuel B. Holabird, quartermaster general of the army. Too young to enter the army at the outbreak of the Civil War, he finished high school at St. Paul, spent two years at West Point and then, like so many of his generation, cast in his lot with Chicago, where he studied in various architects' offices between 1875 and 1880.

I was a boy in Chicago when the first skyscrapers rose. I knew most of the architects and engineers who devised and erected them, and served as a cub under some of them. I come of a family of builders, one of five brothers who have designed and built a vast number of skyscrapers throughout the United States and Canada, and indeed, a few in the Orient and some in Cuba. Three of the five were pupils of Daniel H. Burnham —"Uncle Dan," as he was affectionately called—one of the fathers of the skyscraper and master builders of America. One brother was for twenty-five years with the George A. Fuller Company, for many of those years its president. Another brother founded the Thompson-Starrett Company. These two competitive firms pioneered the new profession of builder and between them have constructed more skyscrapers, probably, than any other half dozen building organizations. A third brother, with Ernest A. Van Vleck, founded the architectural firm of Starrett & Van Vleck. All three concerns continue to be leaders in the industry, though no Starrett remains in any. My brother Paul, former head of the Fuller Company, and I, with Andrew J. Eken, whose family name is so much like my mother's as to suggest a distant relationship, now form a fourth company—Starrett Brothers, builders, of New York; while Ralph, at the head of another oncoming generation of Starrett builders in Chicago, has the Starrett Building Company, in which we are also interested. And as if to preserve an unbroken family tradition, both our sisters married builders, who have given impetus to that tradition by the prominent structures they have erected on the Pacific Coast.

The family tradition has it that my father's father was a builder, and his father before him, in the country around Pittsburgh and Allegheny, Pennsylvania. They were farming and preaching folk of Scotch origin, skilled carpenters, and perhaps stonemasons, who farmed in season, preached, and taught school. My father, whose name I bear, while he had a skilful hand and an appreciation of building, took to the ministry, and the close of the Civil War found him in charge of a frontier Presbyterian church in Lawrence, Kansas, when Quantrell and his men sacked the town and massacred most of its male population. There he had taken his schoolteacher bride, Helen Martha Ekin, and there the family of seven was reared, in the simplicity and hardihood of those early Kansas days, where the minister was too often paid in supplies, and even the sight of money was sometimes a rarity.

Among my father's first acts was to start building a church, still standing, I believe. He interested himself in the founding of Kansas University and had to do with the first university building. Then he started to build his own house, a solid, adequate structure of native stone with considerable effort at design of his own creation. Those were the days in that country when men largely built their own houses, and much of the work was done with his own hands and with the juvenile assistance of my elder brothers. Every part of it reflected his care and thought in arrangement and design. Up to a few years ago the old house was standing and perhaps still remains. Like many another, it passed through its vicissitudes. For a while it was a hospital; then, in the slump of a Kansas depression, it became a tenement, sheltering several families; then it lay vacant for a long while with leaky roof and doors and shutters awry, the windows shattered and its dilapidation almost complete. I went to see the old place just after the close of the World War when some duty in connection with my military service took me through Lawrence. It was indeed a depressing sight, and I was about to turn away when I happened to notice that, in spite of much broken glass, the transom over the front door remained undamaged. On looking more closely, I was amazed to see that it was the original glass, inscribed in beautiful but simple lettering by my father's hand. One of my elder brothers had once told of seeing him do the work; how he prepared the beeswax and carefully spread it on the glass and then, after cutting the letters in the wax, had with acid, somewhere obtained, etched the motto of his household:

> "Into whatsoever house ye enter, first say,
> Peace be to this house."

Needless to say, I made arrangements to secure this cherished piece of glass, and to-day it adorns my own home in Madison, New Jersey, a worthy blessing for any household, great or small.

Interested explorers in the field of eugenics may speculate upon the circumstances of our family origin. To them will be left these matters of deduction, although all of us, sons and daughters, must give the credit of any successes we may have had to our far-seeing and talented father and mother, whose lot was cast in a pioneer frontier town and who, in spite of the handicap of remote location and straitened finances, reared a family of seven, providing us all with good educations.

In the early 80's the family removed from Lawrence to a suburb of Chicago, and my father and mother together took up educational and literary work through the medium of a little paper called *The Weekly Magazine,* which they started in Chicago. Now, it was in Chicago that the skyscraper had its inception, and it was this family of boys whose lives were to be so influenced by the then unimagined skyscraper development. Some of them were destined to take a leading part in the inception, and, indeed, some of them were responsible for at least a part of the pioneer development of the skyscraper of to-day. Prepared by family tradition to be builders, educated to be skilful of hand and self-reliant, some of them endowed at least with talent for drawing, design and organization, and all living in a family atmosphere of creative imagination and constructive effort, it was not unnatural that they should have turned to construction as their life calling.

The oldest boy, Theodore, was forced by family necessities to leave Lake Forest University before he was graduated. A talent for drawing and his builder's inheritance sent him into an architect's office, then into another—that of Burnham & Root. The second boy, Paul, went to New Mexico to ward off incipient tuberculosis. When he improved, Theodore found him a place with Burnham & Root, where he soon rose to a position of responsibility, eventually becoming Mr. Burnham's representative in a number of his large operations in Eastern cities. Later he joined the George A. Fuller Company and, as stated elsewhere, eventually became president of that concern. Ralph, the third boy, worked for a time in a hardware store, then in a bank; then Theodore, graduated from Burnham & Root's and setting up as a builder, was joined by Ralph. Theodore afterward joined the Fuller organization in the East. The fourth boy, Goldwin, alone finished college, being graduated as a mechanical engineer at the University of Michigan and taking his extraordinary talent for design and drawing directly into Uncle Dan's office. I, the fifth son, had to leave Michigan at the end of my second year, worked for two years in a wholesale grocery house, premeditatedly for business experience on the advice of Theodore, and then joined him with the Fuller Company, beginning as an office boy. In later years, Lake Forest conferred on Theodore the degree of Bachelor of Arts, and the University of Michigan gave me the degree of Bachelor of Science in Civil Engineering.

My brothers were Mr. Burnham's protegés, and all of us are eternally in his debt for the inspiration he furnished, for we all came closely in contact with him in one way and another. However, it was not alone for this reason that I think of Uncle Dan Burnham as the greatest of the pioneer builders. He had a forceful, if austere, personality, and his vision was as practical as it was far-reaching. "Make no little plans," he counselled in 1907. "They have no magic to stir men's blood and probably themselves will not be realized. Make big plans; aim high in hope and work, remembering that a noble, logical diagram once recorded will never die, but long after we are gone will be a living thing, asserting itself with ever-growing insistency. Remember that our sons and grandsons are going to do things that would stagger us. Let your watchword be order and your beacon beauty."

The big and noble plans he made for Chicago, Washington, San Francisco, and Cleveland in particular and these United States in general, true to his prophecy, are living things to-day, sixteen years after his death, asserting themselves with ever-growing insistency. My brother Goldwin used to carry Uncle Dan's lantern slides as he went about that raw and crude young city, preaching Chicago Beautiful to any handful in church or schoolroom or lodge hall that would listen to him. Those dreams now are taking form in Wacker Drive, the lake front, the parks and boulevards and a dawning civic centre.

In Washington, he, more than any one else, rescued L'Enfant's noble city plan from oblivion and restored it to health. He alone persuaded Alexander Cassatt to relinquish the franchise on the Mall, that essential avenue in L'Enfant's plan, stretching from the Capitol to the Monument, that an indifferent people had granted to the Pennsylvania Railroad, and he designed the new Washington Union Station, the building of which I superintended for the Thompson-Starrett Company. In San Francisco he built a shanty on Twin Peaks from which he studied a city plan for the Mistress of the Golden Gate. Cleveland called him, and Manila and Baguio, the summer capital of the Philippines. But it was as constructor-in-chief of the Chicago World's Fair that he left his mark indelibly on the nation at large.

CHAPTER II

EARLY HISTORY

WE Americans have been builders from earliest colonial days, and indeed, that colonial period gave us some of our most beautiful and enduring architectural styles. As colonial builders, we were aristocrats and our structures of that period reflect it. With our independence, it must be admitted that a banality set in which is shown in the vast majority of structures built during the first seventy-five years of the nineteenth century. Yet in the metropolitan structures of that period, we can discern the groping of those hard-headed, practical people for the thing we finally attained—the skyscraper. Here and there throughout our large cities one may find even now examples of those banal old buildings, made with cast-iron fronts, too often adorned with clumsy and meaningless ornament. But the structures had a meaning. They were the gropings for escape from the thick masonry walls which are necessary for height unless those walls are relieved of their loads by metal columns. The torch of classical design and architectural beauty was often made to flicker low by blasts of this crudity; but it had its purpose, however obscure, and it is with pleasure that we remember a few great souls who resisted the storm, yet gleaned inspiration from its obscure and uncomprehended purpose.

To my eyes, the most beautiful architectural creation of all time was conceived and erected during this period—the Capitol at Washington; and its crowning glory, the great dome, was made possible by the frantic excursions of these cast-iron builders—the dome is of cast-iron.

But what has all of this talk about cast-iron to do with the skyscrapers? The answer is that the columns of the first sky-scraper were of cast-iron; and there is no doubt that the prior use of this material in fronts and for miscellaneous structural members had shown how these columns should be made.

Now the modern skyscraper is a great complexity of ma-terials and things, and one who would understand it must know something of these materials and things and their his-tory and origins. And here again, unless we take a starting point, we are in a bewilderment that leads nowhere.

The baseline, to use an engineering term, from which to measure the history and development of the skyscraper is the Centennial Exposition in Philadelphia in 1876, for, while there were no skyscrapers or even structures that remotely re-sembled them at that exposition, there were in the construc-tion of the buildings and in the exhibitions many of the seeds that then and there took root under the quickening impulse that the exposition furnished. Construction itself, however, was secondary, and these seeds were unrecognized at this time, for the skyscraper had not yet arrived.

In order to view our progress since the year 1876, it may be helpful to divide the time into three roughly equal periods, for the advancement in the art of modern construction falls naturally into those periods, each period having a distinct characteristic and each being the logical outcome of its prede-cessor in the production of the art and science of building as we know it to-day.

The first period is from the Centennial to the World's Fair in Chicago in 1893. It seems literally that almost everything modern in human advancement commenced with that Cen-tennial, and certainly modern construction can date its genesis from about that time.

A study of the records of the Centennial shows how un-important the structures themselves were considered to be as compared with their contents. Mere shelters they were, adorned with bizarre, jigsaw ornaments. Such structural metal as was used is spoken of in the reports as "iron"—perhaps much of it was wrought iron. The report of the form of construction was in what now seems quaint engineering language, but these structures were no mean undertakings in engineering. After giving certain information as to the height and size of the buildings, the chief engineer speaks of the iron columns—note the word—but the curtain wall had not dawned, for the masonry walls went up only seven feet from the ground, then windows, and then that super-abomination, —galvanised iron cladding on wood.

The Committee on Grounds, Plans and Buildings took a harsh and indeed hopeless view of what should have been one of their main concerns, for they say in effect in their report that design should be eschewed, utilitarianism emphasized, with not one penny for anything but utility. They seemed fairly to gasp when beauty was even mentioned.

The architects seemed wholly to have failed to solve an architectural problem and engineers were called in. Did they fail, or was it the failure of the Board to comprehend architecture? Perhaps the Board, in setting limits so narrow and so unimaginative that no architect could in self-respect comply, was in part the cause of the expressed despair.

The whole temper of the Centennial must be realized as an almost frenzied appeal to a triumphant and material northern manufacturing population to have humanity view the era of unbounded prosperity and progress that the dawn of the events they visioned would usher in. There is a humorous irony in the report of the Director General who, in his scant reference to the buildings, remarks that "the state exhibits were housed in ugly and inappropriate structures"—a sort of vicious kick at a particularly ugly orphan child of a family that had given little else than trouble in the scheme of things as the management saw it.

Machinery Hall, one of the principal buildings of the Centennial.

Sixteen years later the World's Fair Management prepared its Machinery Hall with a skeleton steel frame, a form of construction almost wholly unknown in the earlier Exposition.

So it was that to the engineer fell the lot of arranging and designing the architecture. The unlovely results spoke for themselves, although from the layout of the grounds one must almost suspect that somehow an architect had broken through, and, perhaps in disguise, laid a mantle of good arrangement over the grounds and the placing of the buildings.

It was in 1876 that Charles F. McKim, William R. Mead, and Stanford White made their celebrated pilgrimage to New England, to Marblehead, Salem, Newburyport and Portsmouth, seeking out and making drawings of the best colonial work. Out of that excursion came the classical renaissance in America, but much too late to influence the Centennial.

To me, the most impressive gauge of our fifty years of progress is the fact that, in that exposition of 1876, electricity played practically no part. It is true that the telegraph was used as a means of communication, for the reports tell us of the great convenience it was in communicating with Philadelphia. No less than forty-nine wires were in use in the grounds, and the records are constantly referring to the great saving of time effected thereby. But there were no electric lights or even hints of the possibility of such means of lighting. The piping of the grounds for gas was among the major problems, and the management points with pride to the sufficiency of its street and building illumination by the ample equipment of gas lamp-posts. General night illumination and night display, however, had no place in the plans of the Centennial Exposition.

We, of course, know that one of the curiosities of the Centennial was the telephone, by means of which the voice could be transmitted several miles; but, like the electric lamp, it was a curiosity, a thing in the realm of speculative scientific apparatus with as yet no practical application. Machinery was the thing, and Machinery Hall was one of the principal buildings. The question of power was one of the baffling problems, and

One of the great buildings of the Chicago World's Fair, still regarded as a classic.

The beautiful Court of Honor, Chicago World's Fair, where superb artistic effect took precedence over the banal utilitarianism of the Centennial.

In 1876 this was the last word in a freight handling terminal, as extolled by the management of the Centennial, which this terminal was built to serve.

much is said about the munificent donation by Mr. Corliss of the large engine which stood in the centre of Machinery Hall. Elaborate systems of shafting, both in tunnels and by overhead devices, served to transmit that power to all parts of the building. This mighty Corliss engine is described as a double cylinder compound of no less than fourteen hundred horsepower, supplied with steam from the adjacent Corliss boiler house. The World's Fair at Chicago, sixteen years later, required, in all, over one hundred thousand horse-power, and electric transmission had largely superseded the shafting and belts of the Centennial. Such primitive and experimental dynamos as may have been driven from the shafting were, like the telephone, mere scientific playthings.

Portland cement, while known and in some general use in concrete and for mortar for brick work below grade, was imported from England, but was hardly used in the masonry walls of the superstructure. It is recorded that specimens of reinforced concrete were exhibited at the Centennial as a sort of curiosity in the field of speculative possibilities, but as yet unrecognized for any general use, although a few obscure but daring leaders had commenced to catch the vision. Reinforced concrete, as the reader must know, is concrete in which steel bars or heavy steel mesh is embedded. This steel is first put in position, the concrete poured around and over it and securely tamped in place so as to embed the steel completely. When the concrete hardens or "sets," we have a material of tremendous strength, the great compression strength of the concrete supplemented by the great tensile strength of the steel.

But we were at that time well into the era of bridge building, and rolling mills were even then establishing their standard shapes which afterward were to become the foundation of our skeleton steel design for skyscrapers. Already some long and beautiful steel bridges had been built and the science of civil engineering, as we know it to-day, was fast establishing

From *"The American Architecture of Today,"* by G. H. Edgell.

MICHIGAN STATE BUILDING
How a State building looked, Centennial Exposition of 1876.

MASSACHUSETTS STATE BUILDING
How a State building looked under the skilful direction of the committee of architects, World's Fair in 1892-3.

its basic data. Railroad rails were being produced in large quantities by the already flourishing steel mills, and even these rails were being seized upon by bridge and structural designers for use in foundations and for lintels over wide openings.

Such was the baseline and the background.

Almost unlimited space could be taken in reviewing those Centennial structures and pointing out their quaint and now obsolete characteristics, but it is with the iron framework that we are concerned here. The materials were wrought and cast-iron, probably the greatest structural use of iron ever made in one building construction enterprise until then. Except for domes and towers carrying no live loads, none of the buildings exceeded three floors in height; nevertheless, the seeds of the skyscraper were there. To trace its dim beginnings, however, it is necessary to drop back to the year 1853.

In that year the building of Harper & Brothers, pioneer publishers of New York, burned with a loss of $1,500,000, said to have been the greatest fire loss ever suffered by one American firm up to that time. The Harpers naturally sought to avoid another such disaster and, in 1854, erected in Franklin Square the first so-called fire-proof building of any magnitude in the country. No building is utterly fire-proof, of course, and this was not even so by modern standards, but it was fire-resistant to a point beyond anything that preceded it.

Wrought-iron floor beams were the novelty of the Harper Building. Cast-iron beams had been used occasionally in Europe in efforts at fire-proof factory construction. I cannot fix the first appearance of the much lighter, tougher wrought-iron beams, but they were rolled in America for the first time in 1854 at the Trenton, New Jersey, ironworks of which Peter Cooper, the philanthropist, was the principal owner. The first beams were intended for the Cooper Union, on which work had started in 1852. The Trenton works turned

out the first beams only after two years of costly and troublesome experimentation, and then Cooper waived the first lot to
the Harpers, and the new building of the publishers became
the first to employ wrought-iron beams set in masonry walls
as lateral supports. It was only six stories and contained no
single principle of the skyscraper, yet those wrought-iron
beams were the first faint foreshadowings of the Woolworth
Tower.

As land values and taxes rise, owners must get an increased
return from their properties. This may be done either by increasing rentals or by adding to the rental space of the buildings, or both. Rentals cannot be increased beyond a competitive point; higher than that, renters are driven elsewhere.
Rental space can be increased only by adding stories, and
there the property owner was stopped by the limitations of
the human leg muscles.

Before the invention of the elevator, six stories was the
practicable limit of commercial building. At any time in the
past two thousand years, builders could have erected masonry
structures higher than that, but even the sturdy calves of our
unpampered forefathers balked at climbing more than six
stories, and the rental value of floors above the third fell off
in more than an arithmetical ratio.

In 1859, the new Fifth Avenue Hotel, six stories high, was
opened. It was the last word in modernity and magnificence,
the brightest jewel of its diadem the first passenger elevator
ever built, called by Otis Tufts, of Boston, the builder, a "vertical screw railway." The cab was set on a screw shaft, propelled
upward by a steam engine revolving the shaft, and checked
hydraulically on its descent. Thus it moved as a great nut
would move on a bolt held on bearings, the nut travelling
along the bolt as it revolved. Early in 1860, Tufts installed
another in the Continental Hotel, the pride of Philadelphia.
Each cost $25,000 to build, was cumbrous and exasperatingly

slow, and no more were built; but the elevator age had dawned. In 1866, the first suspended elevator, a steam hoist, was installed in the St. James Hotel, in New York, and two years later, the original Equitable Life Assurance Society building on lower Broadway was erected, the first office building in the world to contain a passenger elevator.

Now buildings could be and immediately were carried to ten stories. Elevator accidents were frequent, but the timorous could walk if they liked. The higher the building, however, the heavier became the lower walls, until the upward thrust of the sky-line encountered another stop clause. Masonry structures of ten stories and more demanded lower walls of such fortresslike thickness and sparse window vents that the ground floor space, most valuable of all, was devoured and the sunlight all but excluded.

In their efforts to lighten these walls without weakening them, architects began to build cast-iron into the brick and masonry. Cast-iron is as brittle as a stove lid, but it has enormous compression strength. The outcome was the heavily ornamented hollow cast-iron front, molded to counterfeit masonry, so common between 1860 and 1880, and of which many specimens survive. It was an excrescence, but another step in the direction of the skyscraper. The Harper Building, torn down in 1925 when the publishers moved up-town, was a notable example. A later fine example was the A. T. Stewart—later the John Wanamaker—store.

Another partial answer to the problem was to reduce the dead load to be carried by the walls. A minor step in this direction was the substitution of hollow cast-iron columns, light and compact, for masonry in interior walls. Most of the dead weight, however, was in the floors. Brick arches, the floor construction practice of the day, were fearfully heavy. Moreover, the lower flanges of the iron beams were exposed to the direct attack of fire. Steel begins to lose its strength at six hundred

The first steel skeleton skyscraper. The Home Insurance Building, Chicago, designed by William Le-Baron Jenney. Originally ten stories high, the two top stories were added later.

The Westminster Hotel in Boston, which was the subject of legislation that established the right of the city of Boston to make arbitrary height limitations. The city ordered the top story removed, which accounts for the roof line. (See text p. 102.)

WANAMAKER'S STORE IN NEW YORK

The exterior walls are cast-iron shells filled with masonry, thoroughly fireproof according to the best traditions of the early 70's

degrees, wrought-iron at about eight hundred degrees, and much higher temperatures are inevitable in a serious fire.

Balthaser Kreischer, a New York manufacturer of fire brick, found the solution both of dead weight and of fire protection in the hollow tile, which he patented in 1871. The flat-arch, hollow-tile floor laid between the iron floor beams weighed only one-fourth as much as the brick arch. Moreover, by this method of construction, he insulated the exposed iron beams with a highly fire-resistant surface and chambers of air, for in his patents the "skew-backs," or tiles that rest directly against the beams, were so designed that flanges of tile reached down and encased the lower flanges of the beams.

Buildings went a little higher again, but now they began to encounter difficulties below ground. As far back as we know anything of building, foundations had been one continuous bed of masonry in solid ground, or wooden piles in wet, unstable soils. Foundations now are a science; in 1880 they were a practice that had not advanced appreciably in thousands of years.

Here the scene shifts to Chicago, where the foundation problem was particularly acute. The city lies a few feet above Lake Michigan on a bed of muck. The ground water level is only ten to fifteen feet below the surface and any attempt at excavation resulted in an immediate battle with water and caving soil. The great fire of 1871 had destroyed the business district utterly, and as the city rebuilt, it did so on a greater, more modern scale. As great weights were imposed on old-style, continuous foundations resting in this soggy soil, there was trouble. The Federal Building, completed in 1880 at a cost of $5,000,000, settled so badly that it was condemned and razed after eighteen years. Frederick Baumann, a local architect, seems first to have suggested an independent foundation for each column, making for more uniform settling.

The pioneer example of this foundation practice was the ten-story Montauk Building, designed by Burnham & Root and erected in 1881. Each pier of this masonry structure, exterior and interior, rested on its own foundation of stepped stones based on layers of concrete eighteen inches thick, the whole designed on a trial-and-error formula of so much foundation spread to so much weight imposed on the pier. The column loads were not so great in those days; perhaps two hundred tons on a single footing was unheard of. On this account the "spreading" for a footing need not have been over large, even though those early engineers knew less than we know now. The masonry pyramids so nearly filled the basement of the Montauk Building that the boilers and engine room had to be located at grade in a court behind the elevator shaft, and the Montauk, like all Chicago office buildings until a later date, really had no basement.

In this early period, Burnham & Root designed the sixteen-story Monadnock Building, the highest that burden-bearing masonry walls ever were carried, I believe. At the basement level the walls are nearly fifteen feet thick.

Spread footings were not new. The foundations of the great cathedrals of the Renaissance were designed by the great engineers of that day, and many of them were amazing accomplishments. But they were based on meagre knowledge and even more meagre facilities. Those early Chicago engineers were just emerging from the foundation traditions of the Renaissance and their work naturally followed the early precedents. Finally, within the limited requirements of all problems then to be met, that Renaissance engineering was generally correct.

It was the ushering in of the tremendous, concentrated loads that gave impetus to our modern science of foundations and brought about the concrete pile, the steel tube and the pneumatic caisson, not to mention the now almost indispensable steel sheet piling. These are the devices of to-day, the complete answer to the requirements that tremendous column loads impose. Deep foundations are only deep because they must first conquer soft, soggy soils and carry their loads to bed rock or hard-pan.

Yet we must not get sentimental about the ancients and their engineering, however much we admire the great architecture they served. The Leaning Tower of Pisa was originally an engineering error. Unstably built, it settled when about two-thirds completed, and for a while was abandoned, awaiting its final collapse. It came to equilibrium through no engineering skill, but because of the uncalculated compression of the soil. After long standing uncompleted, an ingenious theory of its form took the public imagination, and it was finally finished as we see it. Dozens of other examples could be cited. The Campanile of Venice toppled and fell within our memory, due to faulty foundations. St. Paul's in London has continued to settle ever since it was finished over one hundred years ago, its condition in the past decade having become so alarming as to cause the great edifice to be closed for a period recently until the dome could be braced and strapped together. Similarly, the foundations of Westminster Abbey have to receive constant attention. It is only through knowledge of modern engineering that we are able to correct the shortcomings of those old engineers. Who shall say how many of the fabled great works of old succumbed to their faulty foundations? Certainly, the oft-expressed theory that the Colossus of Rhodes fell by reason of unstable foundations is the most plausible explanation of the disappearance of that great world wonder.

CHAPTER III

THE FIRST SKYSCRAPER

BEFORE some architect could attempt to carry masonry walls even higher, the skyscraper appeared. In the fall of 1883, W. L. B. Jenney was commissioned by the Home Insurance Company of New York to design a Chicago office building for them. Others had built cast-iron into their masonry walls and piers and used wrought-iron floor beams, but Jenney went a long and daring step farther. He actually carried out what no one ever had done in theory or practice before—took the dead load off his walls and placed it on a skeleton framework of iron concealed inside the masonry—cast-iron columns and wrought-iron I beams, bolting the beams to the columns with angle-iron brackets.

When the framework had reached the sixth floor, a letter came to Mr. Jenney from the Carnegie-Phipps Steel Company of Pittsburgh. It stated that they now were rolling Bessemer steel beams and asked permission to substitute these for wrought-iron beams on the remaining floors. Jenney agreed, and the resultant shipment was the first ever made of structural steel, in the modern sense. The columns continued to be cast-iron, however, since plates and angles of steel, of which the later steel columns were built up, had not yet been rolled.

This Home Insurance Building, the first of all skyscrapers, still stands at La Salle and Adams Streets; originally ten stories, two more floors were added later. It was started May 1, 1884, and finished in the fall of 1885.

It is true, however, that L. S. Buffington, a young architect of Minneapolis, had dreamed of skeleton steel structures as early as 1880. His inspiration was gained from the speculations of a French architect, LeDuc, who had years before, in a discourse on architecture, written: "A practical architect might not unnaturally conceive the idea of erecting a vast edifice whose frame should be entirely of iron, enclosing that frame and preserving it by means of a casing of stone." Pursuant to the inspiration that this reading gave him, Mr. Buffington set about conceiving multi-storied structures. He dreamed of buildings twenty, thirty, fifty, and even a hundred stories high, and made fantastic sketches. These dream buildings he christened, "cloud-scrapers." He even went so far as to make the engineering calculations as to how heavy the columns might have to be in these buildings of various heights. But for one reason or another, he delayed making any application for patents until about 1887 or 1888. We know, of course, that already the Home Insurance Building had been completed two or three years before. Moreover, it is questionable whether Mr. Buffington could have secured backing to erect any of his dream structures, and whether any of the designs he had made were capable of practical construction. The fact is, regardless of his claim to prior invention, it was Mr. Jenney who put the problem to practical test, and to him belongs the credit, in spite of the commendable excursion of Mr. Buffington into the field of fancy. It is of interest to record that, for several years after skyscrapers commenced to appear in Chicago, Mr. Buffington threatened suit against the owners for infringement of his patents, and it is my recollection that, in one or two cases, he actually started proceedings, but the prior application of the principle by Mr. Jenney largely defeated his case.

The next great step forward came a year later from the office of Burnham & Root. Their twelve-story Rookery Building copied the Jenney skeleton framework, but the founda-

tions pioneered the present steel-grillage design. Instead of
setting the Rookery on a series of bulky stone-and-cement
pyramids, Burnham & Root designed footings of two courses
of railroad steel laid at right angles to each other and embedded in concrete, with steel I beams crossing the upper
courses. John M. Ewen was, at that time, chief engineer for
Burnham & Root, and my eldest brother, Theodore, was then
a draftsman in the office. It was he, I believe, who first sug-
gested this use of railroad rails.

Laymen may find this explanation obscure; but if the reader does not understand how it was done, he will understand
the effect. What was accomplished was a better burden-bearing foundation, occupying only a fraction of the space of the
pyramidal footings and requiring an excavation of as little as
three feet.

Though an advance on the isolated masonry pier, this still
was a complacent acceptance of a floating foundation. When
I was a youngster in Chicago, it was not uncommon for large
buildings to be as much as three or four inches out of plumb,
a condition frequently noticeable in the chatter of the elevators. It was a general practice then to allow for as much as a
foot of settling, and sidewalks were canted upward from the
curb line at as much of an angle as the builder dared, in the
hope that, when the building did settle, the sidewalks would
sink with it to their true plane. The extent of the settling, unfortunately, had to be guessed at.

The real answer to the problem, of course, was to carry the
piers to hard-pan or bed-rock at seventy-five or one hundred
feet and seal out the water to provide a basement; but engineers did not yet know how to combat water and caving
soils, except awkwardly, at prohibitive expense. In recent years
just such foundations have been carried through the muck and
sand and the underlying blue clay to rock or hard-pan under
some of those old buildings — even such a massive structure

as the Masonic Temple—while business went on as usual above. The method is to take the columns one by one, catch them up on girders that span to cribbings placed adjacent, thus forming a temporary straddle between cribs, with the column base dangling over the hole of a now open caisson. These caissons finished to hard-pan are filled with concrete and the dangling base securely embedded. The cribbing is then removed, and the old column rests securely on a new foundation.

And as engineers learned in the early 90's how to tame the ground water, they went back and dug basements under some of these pioneer buildings. A sheath piling first was run down to hold back the sand and water, then the excavation was sealed with a lining of concrete, pitch and five plies of tar paper. A sump was left as an outlet for the ground water constantly thrusting upward and threatening to flood the basement. Pumps keep going year in and out to draw off this water. It is the refinement of this method which now is used in almost all deep basements where water and shifting soils are met.

Though we no longer would build a twelve-story structure on grillages floating on soggy earth and shifting sand, we continue to use this same grillage of steel and concrete as a footing for every pier hole, deep or shallow. What is the necessity in bed-rock, you may ask. Can any mortal-made weight crush the rocky shell of Mother Earth? To an extent, yes. A weight of 1,000 tons resting directly on rock will tend to powder the surface of that rock, however hard; and inasmuch as a variation of a fraction of an inch is to be avoided in foundations, we have to distribute that enormous burden on a spread of steel grillage laid on top of the rock. Moreover, as a practical matter, it is important to have a reasonable spread to a footing on which a column stands to facilitate steel erection, for the bases and grillage are carefully and ac-

Courtesy of Dinwiddie Construction Co.

Courtesy of John Griffiths & Son Co.

curately set level, and the column is bolted to the base before the derrick lets go of it; otherwise it would topple and fall.

The passer-by who stops from a fascination he cannot explain to watch a steam shovel snorting in a hole, imagines that the deeper the hole, the higher the building is to be. This does not follow. We can scrape away two or three feet of earth and run up a fifty-story building or more, if beneath that few feet of earth is bed-rock. If we quarry deeply into the solid rock with air drills and explosives, it is to provide basement and sub-basement space demanded by operating, not engineering, necessities. Contrarily, in swampy ground, we may have to dig one hundred feet to bed-rock to support a ten-story building, and this necessitates pneumatic caissons or some other complex form of foundation construction.

Now, in 1887, one year further along, Holabird & Roche, architects, in collaboration with Purdy & Henderson, bridge engineers, both of which firms still are active, combined and improved upon the achievements of Jenney and of Burnham & Root and designed the fourteen-story Tacoma Building. The outer walls on the two street frontages were purely curtains of brick and terra cotta, carried at each floor by steel spandrel beams attached to cast-iron columns; and here first was seen the startling spectacle of bricklayers beginning to lay walls midway between roof and ground. The Tacoma was the first structure ever built in which any outer wall carried no burden and served no purpose other than ornamentation and the keeping out of wind and weather, which became one of the fundamentals of skyscraper design. The two other walls were masonry and self-supporting. The foundations were the isolated footings with steel members devised by Burnham & Root.

George A. Fuller appears on the scene here as the builder of the Tacoma. He came to Chicago a few years earlier from Worcester, Massachusetts. He was a new type of contractor, pioneering an administrative revolution in construction. Contractors until now usually had been boss carpenters or masons, men of a little capital and foremanship, but generally of no technical education, who executed sub-contracts under the supervision of the architects. This was feasible in small enterprises, but as buildings grew in magnitude architects were overwhelmed with a multiplicity of burdens for which many of them had little training and no aptitude. Fuller raised contracting from a limited trade to both an industry and a profession, visualizing the building problem in its entirety—promotion, finance, engineering, labor and materials; and the architect reverted to his original function of design.

Fuller first was a salesman who sought out property owners and promoted new buildings; secondly, an expert who understood the income possibilities and necessities of office buildings; then a financier who arranged the needful capital and credits; next an engineer competent to oversee every phase of modern building, and lastly, a business executive, buying and assembling materials to the best advantage and commanding a staff of assistants and an army of sub-contractors and laborers. That is the building business as it exists to-day.

Fuller was an engineer, but a builder need not of necessity be an engineer, and it is measurably true that great engineers are not likely to be good builders; the jobs are too unlike. A sound engineering knowledge is of great value to a builder if he first has the other needful qualities, more particularly because an engineer, in his education, learns to observe how things are put together. But the involved calculations necessary to great structures are worked out in advance for the builder by a professional structural engineer. If there were such a thing as a technically educated business manager, he would be the ideal builder, for we are administrators and executives, not specialist technicians. George A. Fuller died in his forty-ninth year, in 1900. He was the victim of his own

tremendous driving power and the demands that the building business often imposes.

There were other pioneers, of course, some very great names
in this skyscraper field, but I did not come into such close contact with them. John Griffiths of Chicago was another like
Fuller, but he never extended his sphere of influence to other
cities as Fuller did. In the East, Marc Eidlitz established a
name for ability and integrity as a builder, to-day made even
more illustrious by his sons, who carry on with increased
vigor the original organization of their father. Norcross in
New England laid the foundations for some of the finest traditions of the modern building industry. And in New York,
Charles T. Wills and John I. Downey left their everlasting
impress on the building profession by the fine structures they
erected. In Philadelphia, John T. Windrim, a pioneer architect of the era of the Centennial and after, gave inspiration to
the sudden skyscraper development of that city and left a
heritage of some of the best structures in Philadelphia. It was
he who designed and supervised the construction of the Pennsylvania Railroad office building, a splendid achievement in
engineering and construction. His son carries on as one of the
country's leading architects, and the skyscrapers to his credit
in Philadelphia are many.

In 1889, the skyscraper evolved into a form the fundamentals of which have come down unchanged in high-building
practice. In that year Burnham & Root designed and built the
Rand-McNally Building, the first skeleton structure of rolled-
steel beams and columns built up of standard bridge-steel
shapes and riveted together. Jenney's Leiter Building, a few
months later, was the first without a single self-supporting
wall, as his Fair Building in 1891 was the first to employ Z
bar columns. Then, in 1890, Burnham & Root designed the
Masonic Temple, twenty-one stories of steel on floating,
spread foundations, the highest building in the world, and

one of the seven wonders thereof for the next few years.

Thus the skyscraper was a quick evolution of some six or seven years, achieved in Chicago and fathered by no one or two men. As Corydon T. Purdy, who himself had an important rôle in its genesis, wrote in 1895:

This reversal of building methods, this change about in the function and use of masonry walls, and the introduction of such new conditions in large buildings, is a real revolution the extent of which hardly can be realized The result is that the constructive side of the problem has reached its most perfect development in Chicago practice. The rapidity and history of its development can be very readily traced in that city. A new idea is tried to a limited extent in one building; a bolder application is attempted in the next; another idea, originating in another office, is worked out the same way. Thus the evolution proceeds and honors are extremely hard to divide.

As compared with its masonry predecessor, the skyscraper was light, airy, sanitary, quieter. Its soar and sweep stirred the imagination; there was prestige in being officed in such a monument. Its structure permitted the shifting of partitions and the subdivision of floors to suit the needs of tenants—impossible in masonry buildings—and its height doubled and tripled the income from a given parcel of ground.

A "skyscraper," said Maitland's American Slang Dictionary in 1891, the earliest known definition, is "a very tall building such as now are being built in Chicago." Literally, a skyscraper is any tall building, but to a builder it implies a steel skeleton incased in a wall that is merely a drapery. There are high masonry buildings and there are some three hundred and fifty reinforced concrete structures of ten stories or more in the United States, the highest a twenty-one story office building in Dayton, Ohio; but though reinforced concrete is as modern a building material as steel, and a sharp contender with it in virtually the whole field of construction, we do not think of a concrete structure when we say skyscraper.

The basis of concrete is Portland cement, which was not

made in commercial quantities in this country until the 80's.
Portland cement, of course, has its important uses in structural steel building, principally as a successor to lime mortar
as a masonry binder. Reinforced concrete is unique in that its
technology was developed by its own scientists and advocates,
and by them applied. Oddly, that development coincided almost exactly with that of structural steel; but where the latter
is wholly American, Europe may claim at least half a share
in the former. On the other side of the Atlantic, the larger
modern buildings usually are of reinforced concrete. Its origins are obscure, but we know that the Ward mansion in New
York was built in 1875 of reinforced concrete slabs and columns and that such were exhibited as curiosities at the Centennial.

Purdy brings us up to the battle between cast-iron, wrought-
iron and steel, which was fought to a conclusion on the field
of the skyscraper. The metallurgical difference between these
three forms of iron is chiefly a matter of carbon content. Cast-
iron is highest in carbon content. It will support enormous
vertical weights, but being highly brittle, cannot be subjected
to cross strains. Thus the columns of the early skeleton buildings were of cast-iron, while the beams were wrought-iron.

What we call steel is really wrought-iron made by a superior process. Bessemer's discovery was simply that, by blowing
air through molten pig iron, he could make a low-carbon
product largely free from the weakening slag common in
ordinary wrought-iron. Steel was a very hard, high-carbon
product associated in the public mind with superiority—the
material of fine swords, razor blades and edged tools; so the
manufacturers appropriated the name "steel" for a wholly
new iron intermediate between wrought and cast, lacking the
essential hardness of true steel, free from the slag of wrought-
iron, yet having the latter's malleability.

Bessemer invented his converter in 1855, but it did not ap-

Where concrete competes with steel. Concrete frame of the Davison-Paxon Store, Atlanta, Ga.; foreground, the Capital Theatre. The concrete girders and floor slabs of the store attach to "party columns" of steel.

Concrete through and through. Not only is the frame of this building concrete, but the walls and architectural design are carried out in the same material.

A concrete frame that will be completely submerged by its exterior masonry. Construction of the Atlanta-Biltmore Hotel, Atlanta, Ga. Schultze & Weaver, Architects.

pear in the United States until the 80's. The textbooks will inform you that steel first was used structurally in this country—likewise pneumatic foundation caissons—in 1874 in the Eads Bridge over the Mississippi at St. Louis; but this was hard, crucible steel, long known and prohibitively expensive for ordinary construction. About 1880, the new Bessemer steel began to be recognized by American engineers as superior to wrought-iron in bridge work, and here the bridge engineers took a hand in the skyscraper. They alone knew anything of the structural limitations of this new iron; by the time of the Home Insurance Building, they had worked out a formula of steel stresses.

To-day we know exactly what a given weight and shape of steel will do; we know it by testing it metallurgically and by ingenious devices, and we assume no calculation to be true until it has been so tested physically. The Olsen compression pump at the Bureau of Standards, for instance, exerts a pressure of 5,000 tons slowly and irresistibly. Other machines have a tremendous compressive strength. Others are so powerful as to pull great rods and bars asunder; and while the force is being applied, measure the diminution of cross-section as the intensity gradually elongates the tested member before it breaks. An impact machine delivers a succession of terrific blows, and with it "break-down" and exhaustion tests are made. Electric recording devices in each case chart the resistance of the column or girder under test.

The chemical composition of steel is regulated to as fine an accuracy as a prescription in a drug store, for, as the material is being processed, samples are taken from practically every step from pig iron to the finished cold product, and chemists in the steel mill laboratories can identify practically any piece of steel from billet to finished product and tell with great accuracy just what the physical properties may be.

Such laboratory verification of mathematical assumptions

A mighty operation in full swing. View of the New York Life Insurance Co. Building, New York, with the stone work well started but many stories of steel still to be set.

California can boast as fine buildings as any in the world. Central Savings Bank, Oakland, Calif.

is a development of the last thirty years. All the bridge engineers could say in the 80's for their stress formulæ was that
they were theoretically correct—and point to the fact that the
bridges stood

It is nearly impossible to appreciate to-day what a daring
thing these men did forty years ago when they ran up buildings to fourteen stories on iron-and-steel skeletons resting on
experimental foundations, on empirical calculations that the
buildings ought to stand. When they did stand and others
even higher and heavier, skeptics were not silenced. What,
they asked, was happening to the steel meanwhile? What would
moisture and this strange new thing, electrolysis, do to it in
time? It was an admitted fact that steel was less resistant to
rust than either cast or wrought-iron. What about wind
strains? Who could say what atomic changes might not occur
in this metal under continuous subjection to such lateral and
vertical stresses? Loads of sixteen thousand pounds to the
square inch were being piled on these columns, where two hundred pounds was the maximum for an equivalent pier of good
brick laid in cement mortar.

Disaster was predicted for years and by men of technical
knowledge. We know now that their fears were baseless, having torn down steel buildings after thirty years and found
their beams and columns outwardly and inwardly unchanged;
but none could prove it then.

The casting and rolling of iron was a great industry in
America. "Iron-master" once was a synonym for power and
wealth. In Pittsburgh the iron makers turned generally to
steel, but elsewhere many of them stuck to the older product
and did not submit tamely to the competition of the new. One
of the advantages of wrought and cast irons was that they
were cheaper and more quickly produced. Bessemer steel was
a luxury product originally, and as increased manufacturing
efficiency brought down the price, wrought and cast irons

kept pace. Wrought-iron rails that sold at $115 a ton in 1837, for example, were quoted at $49.25 in 1880, and $18.25 in 1908.

The big iron founders continued to bid vigorously for business. As late as 1901 we used cast-iron columns in both the Ansonia and the Marie Antoinette hotels in New York, eighteen and twelve stories, respectively, and both splendidly built. The principal disadvantages of cast-iron columns were that they were not continuous and that beams had to be bolted to them. No bolt has the strength of a rivet which, forced red-hot into the punch hole, fills up every interstice and practically becomes an integral part of beam and column. The final disappearance of cast-iron columns in large buildings followed the collapse of a thirteen-story apartment house during construction in New York in 1904. The cast-iron probably was not at fault, but the suspicion was ruinous. In 1877, iron ships universally were of wrought-iron; by 1902, all were being made of steel. A similar revolution took place in building and bridges in the same period.

The open-hearth process, now almost universal, brought the final victory to steel. In the pioneer days of steel construction, sixteen thousand pounds to the square inch was adopted generally as the basic unit of working stress. The better quality steel of the open-hearth process permitted an upward revision of this figure to eighteen thousand pounds. This standard now is in use in nearly all the American cities.

As a measure of the bridge builder's initial contribution to the skyscraper, builders to this day buy all their steel from bridge shops. Steel mills roll only standard and uniform shapes, while the structural steel drawings for a great building call for columns and girders of hundreds of different lengths and strengths. The bridge shops fabricate these beams and columns from standard shapes to the specifications of the drawings, punching the rivet holes, riveting on the lugs and combining shapes into girders and columns. The familiar H column of skyscraper construction is, for instance, built up of various standard shapes riveted together to dimensions specified for the task that particular column is to perform. The mechanics of fabrication have hardly changed in nearly forty years, except that a much larger variety of shapes is rolled nowadays by the mills, and that the steel is of better quality at a lower price.

All the while the elevator manufacturers were keeping one jump ahead of the rising sky-line. The suspended steam elevator was succeeded by the hydraulic and the hydraulic piston types, which were supreme from the late 70's until the early 90's. The electric elevator was slow to appear, because no way could be found to step the load on the motor gently up and down. When this problem was solved, the electric elevator displaced the hydraulic types.

During this period, building materials and appliances in great variety were brought out to meet the needs of this re-created art of building. Terra Cotta, one of the oldest of the ceramic arts, took on a new impetus, yet held its ancient name, and at first could only be made by us in reddish color. The ceramic arts had long ministered to construction, following closely its ever increasing needs. With the advent of the new metropolitan structure, came the demand for new sanitary standards. Enamels and tiles were ushered in, both for sanitary and decorative purposes, and the enamel-surfaced skyscraper was conceived to offset the grime and soot of a smoky city. Heating had come as a corollary to steam engineering, and the science of ventilation by engine-driven fans had been well established even in the early 80's. The advent of electric motors in the late 80's and early 90's had greatly stimulated the use of ventilation and had made possible some of the basic usages still current in the science of ventilation; and last of all, the development of electric illumination and the

High-speed, traction elevators, overhead machinery. The switchboard (left) with controls automatically steps up or down the speed of the car. The motor generators, down centre, are known as the variable voltage control machines, which vary the voltage on the main motors. The cables running over the sheaves of the main machine on the right lift directly on the elevator cars.

Elevator overhead machines showing solenoid brakes, which are held apart against the pressure of heavy springs. In the event of the failure of the electric current, the magnetism leaves the solenoids and the springs bring the heavy jaws of the brakes together, thus stopping the turning of the sheaves and preventing any car movement until the brakes have been released.

telephone had crowned the possibilities of metropolitan building development.

The skyscraper was possible in Chicago because Chicago was young and bold and short on precedent. In New York, with a history of two centuries behind it, the building code was rigid, men were more conservative, and no such experimentation would have been permitted. Only after it had been tried and proved in the Western city—and then hesitatingly—did the New York Building Department approve the plans for the steel-skeleton Tower Building on lower Broadway in 1889. A tablet, set in the lobby before the building was demolished in 1914, to be replaced by one much higher, bore the erroneous statement that it was the first of all skeleton structures.

Courtesy of Victor Mayper.

Demolition of the old Harper Building, New York City. Probably the first "fireproof" building to use metal beams, columns, and girders, and masonry floor arches. Note the ornamental form of the old-fashioned "bow-string" girders left exposed for architectural embellishment. Such construction is now regarded as dangerous, as all modern building laws require structural metal to be encased in fireproofing.

CHAPTER IV

LATER DEVELOPMENTS CREATE AN ARCHITECTURAL STYLE

THE skyscraper has had to submit for forty years to the abuse and patronage of æsthetic critics, many of them architects of note. It was a thing of hideousness and hateful forever, a dry-goods box. Woe unto a people who could produce such an abomination! Going back through the architectural journals, I have been amazed at how rarely a critic sensed that this ugly duckling was a great new art form in embryo, capable of breathless beauties. Even the Woolworth Building, which is a joy forever, was anathema to them, a "five-and-ten cathedral."

The first edition of the New International Encyclopedia, 1904, under the heading of Architecture, dismissed the skyscraper with these words:

> In spite of the radical character of these changes in construction and plan, no sign of any architectural result has appeared. This is in part owing to the purely commercial character of the buildings. They . . . must be as inexpensive as possible in order that the rentals may bear a better proportion to the cost. Hitherto in the history of the world, no architecture of any value has been developed out of any such condition

The further claim may be made for the skyscraper, then, that it did, for the first time in history, produce an architecture of value out of purely commercial conditions.

The conception of the World's Columbian Exposition, known as the World's Fair, in Chicago, arose from a national desire to commemorate the four hundredth anniversary of the discovery of America by Columbus. Originally scheduled for the year 1892, it did not in fact open until 1893. It was a gigantic undertaking, measured even by the standards of to-day,

45

supported by a popular enthusiasm that perhaps was never before approached by our nation. Men of national importance in every walk of life gave of their means and talents, but it is significant that in the great adventure, architecture and building took their proper places of pre-eminence, and the leading figures of the architectural and engineering world were placed in positions of high authority in shaping the destiny of that great undertaking.

It was my privilege as a boy to have lived in Chicago in the very shadow of that great enterprise, and to have known and come in contact with many of the men who created its adequacy and beauty. D. H. Burnham was the great leading figure, but there were others, still great names in the world in which we builders live—Atwood, Root, McKim, Sullivan, Holabird, Coolidge, Day, Post, are names that I remember; and among the engineers, I think of the Shanklands, Henderson, Purdy, Gaiver, Vanderberg, Gray, Weiskopf. These are only a few—personal reminiscences, if you will, with perhaps more important names omitted. It is not my purpose to attempt a galaxy of immortals, but to show the contrast in the point of view brought about by sixteen short years since the directors of that prior exposition had regarded the architect as inadequate in the design of their buildings, and turned the matter over to engineers to rid the management of a troublesome preliminary.

Only sixteen years in time, but centuries in ideals, divided the Philadelphia Centennial of 1876 and the Chicago Exposition. The beauty and splendor of the World's Fair still echoes clearly and no doubt will go down to everlasting posterity as first of all a great architectural and building triumph. Description is beggared by what it all was, and illustrations must recall some of those lovely creations that stirred poets and thrilled the masses with the possibilities of the sheer beauty of architecture when truly applied.

"On the lakeside, a rough, unkempt and tangled stretch of dune and swamp became transmuted into a shimmering dream of loveliness under the magic touch of landscape gardener and architect and artist," Harry Thurston Peck wrote in 1905. "No felicity of language can bring before the eye that never saw them these harmonies which consummate art, brooding lovingly over Nature, evolved into that spectacle of beauty. Not one of the twelve million human beings who set foot within the Court of Honor, the crowning glory of the whole, could fail to be thrilled with a new and poignant sense of what both art and Nature truly mean. The stately colonnades, the graceful arches, the clustered sculptures, the gleaming domes, the endless labyrinth of snowy columns, all diversified by greenery and interlaced by long lagoons of quiet water—here were blended form and color in a symmetrical and radiant purity such as modern eyes, at least, had never looked upon before.

"It was the sheer beauty of its ensemble, rather than the wealth of its exhibits, that made this exposition so remarkably significant in its effect upon American civilization. It revealed to millions of Americans whose lives were colorless and narrow the splendid possibilities of art and the compelling power of the beautiful. The far-reaching influence of the demonstration was not one that could be measured by any formal test. But a study of American conditions will certainly reveal an accelerated appreciation of the graces of life and a quickening of the æsthetic sense throughout the whole decade which followed the creation of what Mr. H. C. Bunner most felicitously designated as the White City."

As Mark Sullivan, who quotes these words in his "Our Times," points out, it was Main Street that went to the fair. The "best people" had a slight disdain for it. And as Mr. Sullivan adds, "Not merely for a decade but unto this day the Chicago World's Fair left its mark in all the fruits of an immense stimulation of the American mind. It took chromo art from the walls of American parlors; to it more than one city owes the greater beauty of its avenues, parks and ornaments; thousands of American homes the greater beauty of their architecture, furniture and decorations; millions of Americans their greater appreciation of beauty and their greater opportunity to enjoy it."

"It is difficult now to realize," Charles Moore wrote in his life of Burnham, "the change that came over American architecture as the result of the Chicago Fair. Nor is it possible to estimate the relative values of the influences that were at work to bring about this change. From the East came men like Charles Eliot Norton, who had been preaching the gospel of the saving remnant and who saw a vision. If such things were possible in Chicago, there was yet hope for a country steeped in commercialism, crude in manners and brutal in the use of sudden wealth. He rejoiced that he had lived to see the day in his own beloved country when architect, sculptor, painter, and landscape architect could be brought together to produce results that recalled, in spirit at least, the triumphs of the Middle Ages."

As such a session of architects, sculptors, painters, and landscape men, gathered to create the Columbian Exposition, broke up one afternoon in 1891, Augustus Saint-Gaudens, who had sat silent in a corner during the entire day, leaped up, grasped Burnham's two hands and demanded:

"Look here! Do you realize that this is the greatest meeting of artists since the fifteenth century?"

The wonders of the exhibitions, of course, kept pace. The gap of sixteen years since the Centennial was almost like the gap of a hundred years that had preceded the earlier event. Supremely acclaimed by the exposition were the science of electricity and the science of structural engineering, the one almost unknown and the other but dimly understood in the

Centennial, now come to be two essential cornerstones of all constructional development. "What hath God wrought!" was the message from Queen Victoria at the opening of the Atlantic Cable in 1858. The exclamation seemed wholly inadequate to those who beheld the structural accomplishments and architectural beauties of the Columbian Exposition.

This second period from 1892, the year of the Chicago World's Fair, until 1914, the outbreak of the World War, is of a very different character from that earlier one. It was in this period that the science and art of building found themselves. Yes, found themselves hugely and magnificently, for they consolidated in that period of twenty-two years all of their own vast accumulations. It was a period of gigantic undertakings and vast construction projects. Adjectives and superlatives unbounded pass in swift array before the mind's eye in the futile attempt to measure and describe all that happened in that brief period, but these outstanding accomplishments proclaim themselves. They were the great improvement in structural steel and the full understanding of its properties and possibilities; the scientific development and application of structural, reinforced concrete; and the high development and perfection of the passenger elevator. Other developments, of course, took their important places, but those three were virtually new, and the elevator, combined with either structural steel or structural concrete, absolutely necessary to any progress in modern skyscraper development as we know it.

Buildings, everywhere buildings! Cities rebuilt and wastes reclaimed. We stand in the midst of our work and try to survey it, and that is all quite impossible at this time. The impetus of the former period took on new acceleration. Volume of construction mounted by geometric progression. All over the country it was rightly the gauge of material progress, and cities vied with each other in their towering structures. The oil-created wealth of the Southwest expressed itself in the great

structures of those prairie cities, and the Southeast, the old
South of the Confederacy, felt the warming pulse of its new-
found industrial life, and proceeded to adorn its quaint old
cities with great metropolitan structures. The building of the
Singer, Metropolitan, and Woolworth towers in New York,
each successively enjoying for a brief period the distinction
of being the highest, was answered by Seattle almost immediately by the construction of the L. C. Smith Building in that
city, some forty-two stories high. Bigness and height commenced to cloy the public mind and the statistics of immen-
sity grew tiresome, as the kaleidoscope continued to turn and
reveal the ever bewildering superlatives of height and size.

More important, if less spectacular, was the development of
the science of building, the science of interior arrangement,
and the appropriate use of materials. American architecture
as a thing of utilitarian beauty in the main held true to its
course, although it did have its debauches and no doubt will
continue to have them. But it progressed with the times; indeed, it led as it should, and already has won the prize of in-
dividuality to which I have so often referred.

During this period the science and art of structural concrete came into practical application and established itself as
a great and indispensable factor. Serving particularly the en-
larging industrial life of our nation, and indeed the world, it
has perhaps found its greatest and best application in the huge
industrial plants of the country, although it has also become
a sharp contender with structural steel in practically the entire field of construction.

Like other branches of construction, the early beginnings
of structural concrete are obscured in the swiftness with which
its development occurred. As in attempting to give the names
of those who originated modern building construction, it is difficult to say just who was responsible for the triumphs of modern, reinforced concrete. Certainly, the names of Ransome and

Hennebique would not be omitted from any roster, but having named a few, the true answer must be that the industry developed itself.

Toward the end of the period we are discussing, we commenced to observe the advent of obsolescence in some of the proud structures of yesterday. Here and there tall buildings of from ten to twenty stories, built in the approved manner, commenced to yield to the wrecker to make way for larger and more important buildings. The amazed architect and engineer who had designed a structure supposedly for all time, stood on the sidewalk and saw their works demolished to make way for newer and better things. It was here that we picked our way over the débris and wreckage to see what time had done. First, how good had been the structural frame, and then, how enduring the other materials? The answer was a complete vindication of the science. We had builded better than we knew.

Foreign lands, during this period, had seen and sensed the growing importance that our structures had played in our rising economic life, and sporadic quickening in some of the European cities gave rise to some few remarkable structures. Their tardy observation of the incidence of our work may have been in a large part responsible for their skipping in a measure the structural steel period and addressing a larger part of their effort to reinforced concrete. Foreign engineers and builders have certainly contributed their full share to the development of structural concrete. But the tall metropolitan building still remained a distinctly American product. Oddly enough, while our architects were studying the ancient classics, these foreign architects and engineers were observing our structures here, and by the paradox that seems ever to beset human nature, they showed a tendency to pass over their own great fundamentals of classic design and go in for a bizarre and, to us, sometimes ludicrous architecture in their groping for a tall building solution. These efforts of design were given the all-embracing name of nouveau art. To-day many cities of Europe are over-sprinkled with these grotesque structures, generally built of reinforced concrete, and to that extent everlasting and indestructible. Perhaps a spirit of repentance will cause their eventual disappearance. But in general, the foreigners accepted the form but not the substance of what American structures were intended to accomplish. Ancient laws and ancient usages, ancient vested rights and lethargy seem to have laid heavy hands on the genius behind the idea, and it can be fairly said that, except for its great assistance in the science of reinforced concrete construction, Europe has contributed little to the great major development of modern building construction.

Japan displayed a curious attitude, for it sent its architects and engineers to marvel and copy, and then they went on over to Europe and studied the classics with a devotion and zeal that might have proclaimed Japan their champion. But the practical, stolid architecture and building of England seem to have strangely deflected the Japanese compass, due no doubt to the preponderant English commercial life in Japan. However that may have been, Japan started rebuilding her paper and matchwood cities with those ponderous and clumsy, thick-walled structures that one associates with Threadneedle Street, and while the Japanese avidly pursued reinforced concrete as well, their monumental works are largely of the stolid British type. I speak now of those structures built up to the beginning of the World War,—the period we are discussing.

The third period is that which spans the time from the outbreak of the World War in 1914 to the present. About fourteen years have passed, but it is a fair prediction that posterity may give it the place of most importance in the three considered, because it measures the economic consolidation of all that had gone before, the recasting of values, the survey of

(*Left.*) Formerly Japanese building operations were enclosed with matting on great poles. Such enclosures are fire hazards. The builders explained that they did not wish to offend the eyes of the Emperor by exposing to view their unfinished work. (*Right.*) Native carpenter at work.

Japanese native stone-cutters. Note the workmanship on the granite lantern in the background.

The hod-carrier gives way to the coolie with the scale beam.

From photographs by the author.

accomplishment and, not the least, the review of the errors and mistakes of the vast tide of construction that had surged upward since that remote time in the early 90's when engineers and architects, still living and active, dared to put into mighty structures the dreams and calculations and conclusions of their own creation.

The sudden cataclysm of the World War in 1914 shook the construction industry, as it did everything else. It shuddered and careened, but in a measure righted itself and proceeded to adjust to the novel economic conditions. Economists have rightly hit upon that year, 1914, as the baseline from which to measure all post-war readjustments. Therefore, beyond noting that the commencement of the struggle engendered a certain acceleration in war industries, with their consequent demands on construction, the earlier event was only the zephyr that heralded the tornado of 1917.

It has been previously said that not every engineer is a builder, nor is every builder an engineer. But the work of the builders in the World War has its important bearing on construction development, and consideration of the builder and engineer as one will help in the explanation of the part the skyscraper builders played in that great drama.

And now we have the difficult problem of defining what happened only yesterday when the mighty structure of the construction industry, grown beyond all measuring—powerful but unco-ordinated, teeming with strength, but not knowing itself—fared forth on the troubled waters of national defense.

Engineering had its birth in military activity. The "ingenieur" was the bridge and road and trench and redout builder of the armies of Napoleon and earlier, picked for his ingenuity by the armies of old, hence his name. West Point was founded shortly after the Revolution to teach the science of engineering to officers, and our earliest ventures in engineering schools laid emphasis on their value as an instrument of

(*Left.*) Japanese solution of "skeleton construction." Heavily timbered roof, and walls with movable "shoji," paper-covered sash. Plain wall surfaces constructed of a weaving of bamboo wattles plastered over smoothly with clay mixed with wood fibre. (*Right.*) Japanese workmen take to rivetting.

Both in setting and design, the Imperial Theatre, Tokio, shows the Japanese appreciation of classic architecture.

A corner of the Imperial Palace Grounds, Tokio. Simple native architecture and virile ancient stone construction.

From photographs by the author.

national defense in training engineer officers. And yet the nation was hurled into that mighty struggle knowing the builders and engineers on every hand as individuals, but knowing them not at all as a great potential force, necessary beyond imagination to the war preparation and holding the key to what the nation must first of all have—structures for war preparation.

It was my privilege to serve as an officer in the army at a point where I could well survey the construction aspect of the mighty drama. And it may, therefore, be of some interest here to review my observation of what happened in those early few weeks when the nation, in a frenzy for munitions and men and implements, fairly overlooked its first prime necessity and left construction and its co-ordinate forces to proclaim themselves.

A few hours' inquiry showed that, among other things, the army would require sixteen cantonments, each to house and train about forty thousand men; sixteen national guard camps which, while lightly dismissed, turned out to be about as complicated as the cantonments; some twenty to thirty flying fields, which were in effect great military posts; ammunition storage bases, themselves veritable cities, warehouses at a dozen points, any one of which would exceed in size almost any then existing terminal in the country; and gigantic exportation bases at strategic points along our seaboard. This was for the army only. The navy had its demands which, while not so extensive, were of equal importance. Then there appeared the enlargements of industrial plants that had to be constructed with government funds and under government auspices. And for the organization to handle these things, then only faintly perceived by the army in its anxiety to make ready for war, there were a colonel and two captains in the Construction and Repair branch of the Quartermaster Corps.

There is no intention to make this a humorous arraignment of our government, but an indictment of ourselves as a na-

Owing to earthquake conditions Japanese skyscrapers are limited to 100 feet in height or about eight stories. This large office building in Tokio has a heavy structural steel frame to resist earthquake action. The outside walls are of a hollow common brick, and the exterior surface is a tile about an inch thick in imitation of face-brick.

Transporting a ten-ton steel column by means of bullock teams in Japan. Heavy motorized equipment for this kind of hauling is not yet in use. The light culverts and stone bridges over the canals in the cities of Japan present great difficulties for heavy traffic.

Courtesy of George A. Fuller Co.

tion unprepared. Where had we been? What had we engineers and builders been doing that all of this monstrous situation could possibly be? The truth is we had, with the rest of the nation, dozed off in the lethargy of smug contentment, with calamity fairly shrieking at our gates The lotus sleep, dear to the heart of the pacifist, had all but overtaken us. Those early nightmare days and nights! No offices—no stenographers or facilities, a little stationery filched from wherever we could get it—conferences held in elevators as they plied up and down—meetings frequently held leaning against the marble wainscoting in the corridors of the over-crowded buildings of the War Department in Washington. The one outstanding light in the bewilderment was the instant response of the building and engineering profession. They came to Washington from everywhere—at our request and their own expense. We met, discussed, counselled together, and out of it was evolved that triumphant organization, the Construction Division of the Army. The cantonments were launched amid the stress and clamor of a thousand contending voices and the killing discouragements of a bewildered government. How proudly we record the way in which the men of our calling hurled themselves into the effort. Fortunes were disregarded. The government might take their all if only they should be allowed to proceed with the tremendous tasks, now in part perceived and crying for instant commencement. Men took responsibilities involving millions with no better authority than the word of a construction officer. Into the field moved the hastily organized supervisory forces of the government with the engineers and builders, there to work out together as they built, the details, major and minor, of the cantonments and camps so sorely needed. Wasteful, yes—for we were paying the penalty of impotence and unpreparedness, but honestly and conscientiously done, with little heed to personal consequence. The peace-time order of things had all been reversed. Speed was the prime consideration—cost of secondary importance.

Such was the part of the builder of skyscrapers in the World War, harrowing, anxious months that seemed like years; but they were supreme in adventure.

And now we look about us to see what we are to-day, and perhaps have the audacity to enquire whither we are going. What has been the characteristic of the period from the war until now? More buildings and yet more buildings. The mighty tide of construction was not interrupted, but only diverted to war requirements. And we are apparently continuing the geometric progression of progress and volume. Let us not so much try to measure it as to observe what it is doing. Since the close of the war economists agree that we have been in a great current of economic readjustment, only dimly perceived in its consequences, yet all-embracing and with a portentous destiny.

One singular and significant fact of skyscraper construction here stands out, and that is its power fairly to move centres of cities. Under the old order, the cross roads, vast traffic streams and dense population knew no law but itself, and intensive building submissively remained where density had ordained it should stay. But now it is different. Great structures can actually beckon the trends of population and traffic, and in a measure can compel the shifting of economic centres of gravity. Witness the growth of Cleveland as it shot out Euclid Avenue, abandoning the century-old focus at the Square, and creating in a decade substantially a new city where only yesterday were residences. Washington, our national capital, continues to grow toward its new business settlement northwest of the Treasury and the White House, those far-flung outposts set by former generations to mark the utmost possible limits of growth in that direction.

Dozens of like examples could be cited, for what growing

city is there in this country that does not feel the urge of the
modern skyscraper to readjust its focus to new locations? The
forces that compel these changes are diverse and complex, but
certain it is that our modern construction plays a large part
in the cause—a new economic force in the world.

Our buildings since the war have been all that they were
before, but we are scrutinizing them from an added view-
point. Their economic relation is more clearly observed. In
industry, as never before, the fitting of the building to the
processes of manufacture has taken an enormous impetus.
High costs have forced the minute scrutiny of every foot of
constructed area and volume. The obsolescence of earlier structures has been enormously accelerated by these considerations.
High specialization of industry has had its telling effect, and
never before have we been so called upon for specialized structures. This is true not only of office buildings, but of other
forms of intensive, metropolitan structures. Obsolescence fairly
treads on the heels of early maturity. This is more particularly
true in all forms of intensive residential construction, such as
apartment buildings and hotels, but it is also measurably true
of office buildings. The interesting thing is that the obsolescence does not arise from the form or methods of construction, but from quickened social and business requirements
brought about by demands for more conveniences and better
efficiency. Poor planning in the original instance, antiquated
forms of finish and facilities, antiquated and insufficient
mechanical equipment, and insufficient size—all work for
obsolescence in the older buildings. We almost never hear of
buildings of skyscraper type being torn down or abandoned
through the shift of location, for causes arising out of decreas-
ing property values; the demolitions are almost invariably to
make way for bigger and better structures.

There are notable examples of buildings built even as far
back as the late 90's, and numerous buildings built within the

succeeding ten years, which have held their own with even
the latest and most elaborate of our modern structures. It is
true that these buildings that survive in this way have been
rehabilitated from time to time, new elevators installed, new
and modern plumbing and sanitary facilities introduced, as
much to keep pace with the latest creations as out of any
sheer necessity. The passenger elevator is not unlike the automobile, in that it has been improved from year to year; the
older types being superseded, both because they wear out and
because of the vast improvement in the art of elevator construction. The interesting fact is that, as builders, we in large
measure anticipated these things in the best of our early buildings, and the provision for bettering the equipment of these
skyscrapers was thought of by the original designers.

However, good or bad, the early designers had from the
very outset solved the structural problems for all forms of tall
buildings, and those early solutions in the use of structural
steel still remain the foundation upon which all later designs
have been built. True, we have built higher and ever higher
buildings; girders and trusses of undreamed-of proportions
have been constructed to meet the needs of the vastly complicated skeleton construction that the greater buildings require.
Not only have we learned to build these columns and trusses
but also to handle them at the building site—no small part of
the problem. But the fundamental principles of civil engineering remain the same—truly solved in the earliest instance by
the structural engineers who set about this great problem in
the years succeeding the Centennial.

We speak of poor arrangement—poor planning—as a
cause for obsolescence, and are at once reminded that architecture is not only the art of design, but is also the art of good
arrangement. That lesson of good arrangement is one that is
ever being learned, and one that perhaps will never be fully
and finally solved. Reference has been made to buildings built

in the early 90's, which still hold their own as competitors with the most recent structures. This is largely due to the true understanding of the art of good planning—so well perceived by a few of the early architects of this new form of construction, as keen in their vision for the utilitarian requirements of the structures that they designed, as in the vision of the possibility of the skyscraper itself.

It is almost idle to speculate what further improvement there will be. The speed of elevators seems to have reached its limitations, due to the requirements of the physical comfort of the passengers. Interior lighting seems as nearly perfect as human requirement would ever demand, and certainly plumbing sanitation and the kindred conveniences are well enough understood now to give assurance that no great changes will be made. It is thinkable that steam heating will be superseded by electrical heating. The method of obtaining electric light at the desired point may, in the eyes of the visionary, some day be accomplished by wireless means. But with all of these improvements in mind, the form and good arrangement of the skyscraper, as far as man can see, will be about the same. It will perhaps be higher and undoubtedly cover greater and greater areas. Nearly all of the demolitions that we now see, as I have said before, are the replacing of smaller buildings by larger ones, both in height and in area covered. Buildings covering a whole city block are not now uncommon, and they will continue to grow in number. This arises from the economic situation too well understood to be discussed at length. The operation of these great buildings has become a science in itself, and small units under individual separate managements must yield to great units under unified management. Such is the trend of the times!

CHAPTER V

BUILDING A SKYSCRAPER

IT is a fascinating game, building these great skyscrapers, and to those of us who stay in it year after year it's like strong drink; we get so that we just cannot do without the strenuous activity of it all. And it is a compelling thing, too; a man gets his pride up over it, pride of accomplishment, pride in making good on prediction and forecast. "It can't be done" carries a challenge that the dyed-in-the-wool builder sometimes too eagerly accepts. So our business casualties are high—all too high for the sweat and toil that go into the game.

Building skyscrapers is the nearest peace-time equivalent of war. In fact, the analogy is startling, even to the occasional grim reality of a building accident where maimed bodies, and even death, remind us that we are fighting a war of construction against the forces of nature. And the spirit of the Crusader is there, not only in the daring and courage, but also in the grief that ensues on these occasional terrible accidents, for they do happen in spite of the millions that are spent annually to prevent them. But the analogy to war is the strife against the elements. Foundations are planned away down in the earth alongside of towering skyscrapers already built. Water, quicksand, rock and slimy clays bar our path to bedrock. Traffic rumbles in the crowded highways high above us, and subways, gas and water mains, electric conduits and delicate telephone and signal communications demand that they be not disturbed lest the nerve system of a great city be deranged. Yet we venture down and do it, and from that inaccessible bed-rock or hard-pan we turn back upward, with accurately laid and adequate foundation structures to support the incredible weights that the columns of the skyscraper will impose; for it is not uncommon to have a thousand tons on a single column and a hundred separate columns and footings is not over-many as we build skyscrapers nowadays. All this is done without settlement or movement of so much as an inch, and with accuracy of position that regards the variation of an eighth of an inch as the very limit of allowable error.

Before these things are done, nice calculations have to be made of the weight of the building—just how much loading will be carried on each column; and then the ground is studied and soil conditions analyzed. Not content with all of this, borings are taken and test-pits dug—the reconnaissance of a combat attack on an opponent entrenched from geological eras and having all the mighty forces of nature in alliance to resist and oppose any disturbance of its age-long somnolence.

And with the foundations completed, only the first trenches have been carried, for the superstructure allies itself with gravitation and wind-pressure to resist the accomplishment; the same gravitation that, in its resistance, serves to make stable and everlasting the great skyscrapers that the battle is fought to produce. Rain, snow and sleet repeatedly attack, sometimes with rushing tornadoes, sometimes with the long protracted attrition of rainy seasons or continued blizzard and cold. Even after the steel of the structure is up and the walls well toward completion, as if to make their last violent sortie, blizzards rage in those dizzy heights, numbing the bodies and hands of the intrepid workmen at their tasks away up aloft, while in the streets the busy throngs muffle themselves in warm clothing or gather around cheerful radiators, rejoicing that civilization has so far advanced to mitigate the terrors of an adverse climate. Yet, civilization has done all these things, and away up there where the icy winds sweep unrelentingly, the forge fires of the riveters hiss and glow while numb hands buck up with cold steel dolly-bars, and jack-hammers rattle

The New York Life Insurance Co. Building, New York. The completion of a great stone exterior is not necessarily the completion of the building. The complicated copper work and the great cast-bronze lantern in the tower require many weeks of work by expert metal workers. Cass Gilbert, Architect.

and clatter defiance to the elements. These are the outposts fighting nature back, that nature may be made subservient to our common need. Building skyscrapers epitomizes the warfare and the accomplishment of our progressive civilization.

Even the organization closely parallels the organization of a combatant army, for the building organization must be led by a fearless leader who knows the fight from the ground up, knows the hazards of deep foundations, and the equipment that raises the heavy steel and sets the massive stones one on another, the hoists and derricks, the mixers and chutes, the intricacies of all the complexity of trades that go to make up the completed structures; what they may be made to do, and where making must cease and daring must be curbed; where materials and things come from, and how long it takes to prepare the different kinds; what to allow for contingencies of temporary defeat, and how to consolidate the gains. Ever pressing forward, that leader with his lieutenants and they with their sub-lieutenants plan and do, ever prevailing over inertia, animate and inanimate, until the great operation fairly vibrates with the driving force of the strong personalities that direct the purposes of everything, seen and unseen, that makes for the swift completion of the work in hand.

The obtaining of materials near and far and the administration of all those thousands of operations that go to make up the whole are the major functions of the skyscraper builder. Knowledge of transportation and traffic must be brought to bear that the building may be built from trucks standing in the busy thoroughfares, for here is no ample storage space, but only a meagre handful of material needing constant replenishment—hour to hour existence. Yet it all runs smoothly and on time in accordance with a carefully prepared schedule; the service of supply of this peace-time warfare, the logistics of building, and these men are the soldiers of a great creative effort.

Marshalling this chaos into order is the commanding field-officer of the builder's troops, the job superintendent. The first building on the site will have been his shanty and there he will be found workday, holiday and Sunday, until the architect signs the certificate of completion. He knows how to be vice-president in charge of operations, master mechanic, superintendent, train master and chief despatcher keeping the traffic of a four-track road moving on time through the bottle-neck on a single pair of rails. He may be a jack of all trades and a master of many, but he is no superintendent unless he knows how to organize and how to delegate. Emergencies pound at his door day and night, and he must know the answer without looking in the book. Back of him stands the master builder who has travelled the same road and who has equipped him with everything necessary to the job except the know-how and the driving power; these he must supply himself. A poor superintendent often is felt on the job before the cost sheets and the slipping time schedule reveal his incompetence utterly.

As the young engineers come out of college and enlist as clerks and time-keepers, their goal after a year or two of apprenticeship is to be an assistant superintendent, or, as he is inadequately called in the trade, a job-runner. These assistants are not, as their name implies, the superintendent's lieutenants, but liaison officers between builder, architect and sub-contractors. While the skyscraper gets under way they are in the architect's office asking how this detail is to be built, how that, and carrying the information to the sub-contractors who make their shop drawings therefrom. For each sub-contractor must be prepared to take up his own special task in order, and few items come in stock sizes.

While the steel work is nearing completion away up in the upper reaches of the skyscraper and we break out with our masonry on one of the upper floors, where the substantial

brickwork of the building as it rises above the slender, exposed columns of the lower floors makes the structure look a
bit ludicrous, many things are happening, not only in the
complicated interior of the building, but at the shops, in the
mills, and in fact, far and near throughout our great industrial
centres; for they all minister to the building of a skyscraper.
Organized forethought translates itself into active organization. Here in a manufacturing plant a thousand miles from
the job, an expediter from the builder's office is checking up
on a great blower of special capacity and dimensions, designed
by the ventilating engineers while the job was still merely on
paper, to fit in a certain cramped location away down in the
bowels of the building. It is about ready for shipment and is
needed because the sidewalk beams cannot be set until the
blowers arrive. The steel erectors are about to finish, away up
aloft, and the job superintendent knows that it is important
to have the sidewalk beams completed while the erectors are
still on the job. Moreover, the time schedule calls for it, that
schedule prepared long ago before the wrecking commenced.

In New England, another expediter is arranging for the
loading of the granite, checked off piece by piece, for every
piece has a separate cutting diagram, not only showing each
stone as an integral part of the design, but also showing how
that granite must be cut away on the back to fit securely on
those same sidewalk beams that are giving the man in Detroit so much concern. Inside, plumbers, steamfitters, electricians swarm over the job getting in their piping, for pipes are
everywhere, while down in the depths, now brightly lighted
day and night, sheet metal workers hang great ducts that
twist and turn and dodge pipes and squeeze between girders,
that fresh air in ample volume may be conducted to those
depths by the fan that Detroit is making. Drawings, always
drawings, depict all of this; those same drawings that the engineers were preparing when the site still held the old build-

Heavy masonry foundation avoided by building on
steel stilts. The building fronts on the street at the
high level.

Photograph by Ewing Galloway.

Washington Square, Washington Arch, and the be
ginning of Fifth Avenue, New York. The skyscraper
tower by Helmle & Corbett, Architects.

C. W. & Geo. L. Rapp, Architects.

Night illumination used to glorify architecture on the

Trowbridge & Livingston, Architects

The Equitable Trust Company Building, New York.

ings now demolished. When that granite arrives, it will be set to align perfectly with the masonry of those walls started away up there while yet the stone was being cut. In all of this, and in fact, throughout the whole conduct of the job, the column-centres on the architect's and structural engineer's drawings control all matters of exact location.

Thus, the steel columns must have been set true and plumb; plumbed by a special crew, after the erectors have left the floor and before the riveting on that floor is started, for the erectors only hastily bolt their work as they forge ahead; yet the riveting must follow close behind, for the masonry floor arches are also pressing from below in this race to reach the top. How it all dovetails! One trade following and intermingling with another, yet all in orderly fashion and all in accordance with the schedule. It is the job superintendent who controls this great piece of team work, and as the succeeding trades are marshalled and embark upon their work, the whole job becomes imbued with his driving force, and every one on it senses the quality of leadership that guides it all. Here again the analogy to a combatant army is striking, for we all know that our fighting forces are only as good as the officers who lead them. Yet in the main office, the skill and experience of the master builder control. In spite of much that we see going on in this building game that gives discouragement, and sets men of life-long training and devotion to wondering, building is a great and inspiring calling, and the master builder is still supreme. Quality in building still asserts itself, and our great national pride, the skyscraper, still holds bounteous opportunity for the exercise of the builder's art and forethought.

The spectacular work of excavating, shoring and foundations, particularly where steam shovels are employed and heavy derricks lift great weights, has a fascination for the public to the extent that the crowds of spectators on the side-

The Chicago Temple Building, Chicago; Holabird & Roche, Architects. In commercial centres churches have in many cases been forced into combination with office buildings.

The fulfillment of W. B. Foshay's life ambition —a thirty-two story tower in Minneapolis, with "battered" columns tapering the building like the Washington Monument. Magney & Tusler, Architects.

walk would actually block progress, not to say endanger their
own lives, if they were encouraged to loiter; hence the forbidding fences that the builder puts around the lot as soon as
he can. As the structure rises, steel setting is perhaps even
more spectacular, but here the point of vantage is anywhere
in the street, and the public sees it and has learned to understand it better. Setting of exterior stone work has but a meager
fascination after the spectacular feats of the steel workers, and
brick-laying is almost prosaic, excepting that we like to note
the progress from day to day and admire the speed with which
a great structure is enclosed. A story of brick work a day is a
usual accomplishment for the well-organized builder. Then
follows the glazing, and all of a sudden, the building seems to
stop, for it loses its spectacular interest.

Very different is the situation within those four walls,
where to the builder's eye the major part of the operation is
still to be done. The floor arches have been finished and the
piping for the electrical, plumbing and heating work has all
been done. Then comes the bustle of building partitions, always of fire-proof material accurately laid, with doorways located and minute attention to floor plans, because already the
renting agents have been renting space and there are many
special lay-outs and arrangements upon which leases have already been closed. All the special locations of electrical outlets,
base plugs and telephone outlets, if the lay-out has been carefully made, must be indicated for the guidance of the builder,
and they are taken care of as these masonry partitions are being built.

The plumber's "risers" and piping must be tested before
they are built in. If the building is a tall one, this test is carried on in sections of ten to fifteen floors at a time, thus avoiding too long delay in the partition work Joining and following all of this comes an army of carpenters, metal lathers, sheet
metal workers, marble and tile workers, cement floor finish-

ers, elevator constructors, and what not—all marching in interlocked procession, for the sequence is complex.

In the superintendent's temporary office down on the bridge, daily conferences are held, either the foremen or representatives of the various sub-contractors attending, all intent upon working to the common plan which the superintendent has set, the same plan that was devised in the builder's main office away back when the time schedule was made and the builder planned the execution of it all.

The completion of plastering is the goal of the builder toward the finished work, as the completion of the steel is the goal on exterior. The enormous amount of rubbish and dirt incidental and seemingly necessary to plastering stops all consideration of the finish, such as painting and carpentry where interior woodwork is used, and indeed, even marble and tile work; for, while marble and tile are sometimes set close on the heels of plastering and sometimes even before the plastering is finished, these two items appeal to the visual satisfaction which, in the last analysis, must be the criterion of the builder's art. He may build with the conscience of a saint and the integrity of a trustee, but if in the finished building plaster surfaces are not true and straight, if expensive marble is cracked, tile work chipped, and painting marred, condemnation pursues him.

It is for this reason that the expert builder still clings to craftsmanship, and encourages it, for in spite of the standardization and pre-construction which have inevitably followed high wages and the division of labor, we cannot get away from the necessity of craftsmanship This necessity has prompted the New York Building Congress to establish as one of its most important functions the awarding of craftsmanship certificates and tokens. This custom, although young, has already seized the imagination of the building industry and the architectural profession.

The initial move in the process of awarding these certificates arises when a special board of the Congress selects some notable building under construction where craftsmanship of high order may be expected. A representative of the owner, of the architect, the builder and of labor make a study of the work for a period of several weeks prior to the award. On the appointed day, there is an appropriate ceremony and one craftsman in each trade to be considered is given his diploma, together with a lapel button, not necessarily for the workman of greatest productive effort, nor yet for minute skill, but, as the word implies, for general craftsmanship and high ability. Needless to say, such awards are highly prized by the men and have a significance beyond the mere token presented. It is the effort of the building industry in all its interests to return to that spirit which guided the great structures of old and from which we seem unhappily to be parted in the mad mechanical application of our new-found instrumentalities of building construction.

We Americans always like to think of things in terms of bigness; there is a romantic appeal in it, and into our national pride has somehow been woven the yardstick of bigness. Perhaps that is one of the reasons we are so proud of our structures; they are big, very big, certainly the tallest and certainly the most complex and the most compelling the world has ever seen. They fairly personate the hustle and bustle of our modern accomplishment and postulate our ideal of efficiency, and they are our national pride because they are so completely American So the bigness of the business as a whole we enjoy gasping over. Just think of it—over six billion dollars a year are spent on recorded structures in cities and towns that have official records on such matters. And that is not all, because throughout the length and breadth of the land, in hamlets, on the plains, in the mountains, everywhere there is sure to be building of some sort—always the spontaneous prod-

uct of a virile and progressive people, always the token of a
progressive nation. One enthusiast puts the unrecorded building at another six billion, but no matter—one six billion is
enough to cloy the mind and give sufficient warrant for our
claim that we are the greatest builders the world has ever
seen.

Billions roll off the tongue so easily that to come back to millions seems like a humbling of our thoughts. Yet the spending of
a few paltry millions in a single structure, all within the compass of a year, may still hold an interest; and when one views
it as a great and complicated operation involving skill and daring, with a wealth of adventure and the joy of fulfillment of a
hard task well done, the scale of bigness may again grip the
imagination, and in the story of how it is all done may yet be
held the romance of a triumph no less stirring than the victory
of battle, or the leading of a nation into the paths of peace and
prosperity.

Courtesy of Marc Eidlitz & Son, Inc.

Harkness Memorial, Yale University, New Haven, Conn. A mediæval design of modern
construction. James Gamble Rogers, Architect.

CHAPTER VI

PLANNING AND FORETHOUGHT

ARCHITECTS design buildings and draw plans in interpretation of the requirements of the owner. Engineers design the steel skeletons and foundations in accordance with the requirements of the design. Other engineers design heating, lighting, plumbing and ventilation in accordance with those same architectural requirements. Builders devise ways and means of accomplishing the completed whole. And the lowly owner pays or devises ways and means of payment for it all. Some men or organizations are combinations of two or more of the separate functions, and indeed, of late years, we are observing a few combinations of all of these in one great complexity. As a complexity it is prone to be confusing, but it is of the builder only that I speak. Whatever meed of credit and praise is due to each, the great American public takes the skyscraper builder to its heart; in his ways, his devices and ingenuities, there is the romance and the drama, and it is in him that the breathless interest centres when he comes on the stage. The name of the drama is ever the same, yet ever fresh and exciting; it is the Building of the American Skyscraper.

He starts to demolish, and at once we all take notice. Some eyesore, we are pleased to note, is at last to give way to a great improvement; or, we observe with regret that some cherished landmark has at last yielded to the demand for a bigger and better structure. But if so, "The King is dead; long live the King," wells up in consolation, and in a trice we are lost in admiration of how it is done, and eager to watch the work in its swift progress toward completion.

76

Foundation of the New York Life Insurance Building, New York. The excavation was blasted out of solid rock and is one of the largest rock excavations ever attempted in Manhattan. In the deep basement the rock had to be excavated for over 72 feet below the street level. The rise and fall of the line of solid granite of the Island of Manhattan is clearly shown along the walls.

It all starts away back in an architect's office, sometimes a year or more before the demolition we see, sometimes only a few months; but in any event it has to be worked out on paper beforehand, the floor plan, the design, the engineering of all sorts. The cost has to be estimated as the plans are being drawn; a budget of the cost of the various parts is prepared for the guidance of all concerned. The steel design, all important, must have been completed. Tests of the soil are made, or at least soil conditions fairly well understood to establish foundation designs. Plans for heating, ventilating, plumbing and electrical work are made; in fact, the building has, in a way, already been built on paper before the work we see starts.

After my early apprenticeship as office boy, it was my good fortune to be sent out on one of these operations of sizable proportions as timekeeper, and it was there that I first took part in the application of much about which I had heard. It was the everlasting planning that interested me; not only the architect's planning, for that had already largely been done, although a deal remained to be done, but it was the planning of ways and means—how best to get the old buildings down, how so to conduct the wrecking that excavation could start at a certain time, and, when started, how it could continue at full speed without interruption, and how the succeeding step of foundation building could start; how adjoining buildings were to be held up while our work went on uninterruptedly without danger to those adjoining properties. Planning, everlasting planning—the ways and means are the constant concern of the builder.

Now, a builder, to be any sort of a builder, must work to a time schedule prepared with forethought and out of his experience and ability, and I learned in my first job how this was done and how all-important it is. Time, as well as money, is spent and both must be budgeted; and the drive is always to keep all branches of the work approximately within the

An addition to this building in Detroit made it necessary to replace the columns below the second floor and add piers forty feet deep for the new columns. Temporary trusses were installed and the load of about 500 tons to the column was transferred to temporary piers and columns which were pretested to overload. The new columns were then installed.

Hudson Department Store, Detroit. Making the general cellar excavation after the caissons, columns, and cellar walls had been installed in pits and the steel work erected.

Courtesy of Spencer, White & Prentis, Inc.

money budget and the time schedule. The essence of the building of these great skyscrapers is organized forethought.

That first outside job of mine was on a comparatively simple building, and at that time methods had not been developed as fully as now. We struck a sort of quicksand unexpectedly under several of the column footings, and the assumptions of the bearing value of the soil had to be revised. We knew from the engineer's calculations just what load would be carried on each footing—that was fixed—but the soil being softer than was calculated, the "spread" of the footing had to be increased; that is, it had to be made to cover a larger area so that the weight on each square foot of soil covered by the footing would be less. Thus, under those columns carrying four hundred tons, the soil was calculated to bear four tons to each square foot, and the footings were designed to have an area of a hundred square feet, or about ten feet square. But in this case, the soil was declared by the engineers to be good for only two tons to the square foot; and so the footing under each column where this soft soil occurred was enlarged to a trifle over fourteen feet square, or two hundred square feet, and the desired spread was effected. And here we have a concrete example of a simple foundation problem. Foundation work of all sorts, whether in rock, on hard-pan or of floating type—spread to effect a compressing of soil of known carrying capacity—all must take into consideration the loads that each column will impose. So it was with the underpinning of the adjoining walls; for when an excavation goes deeper than the foundations of a heavy adjoining structure, underpinning must be reckoned with and soil conditions play an important part.

I have seen some wonderful feats of underpinning. Huge buildings, ten to twenty stories and more, caught up on one side and held in place while still deeper and more elaborate foundations for the new adjoining skyscraper were built alongside. And all this was accomplished without so much as a crack showing in the building shored and underpinned. Where this is done with quicksands oozing, and tides rising and lowering, with rains and surface waters clamoring to undermine the work, we have one of the great feats of skill of the builder, the accomplishment is all his and so is the glory. He has here truly devised ways and means—his real function splendidly performed.

After a year or two, in which my timekeeping had taught me much beside timekeeping, I was assigned as superintendent of a sizable structure at the intersection of two busy New York streets, but not before I had had an opportunity as job engineer to run levels and set grades and to lay out the accurate lines to which the column centres were laid. It is pretty work, this field engineering, carried on with meticulous accuracy, if all parts of the building are to fit. In busy cities where land values are measured in millions, and public officials are jealous of encroachment of even small fractions of an inch beyond the building line, where title policies require a nice accuracy of location, and where an error of more than an eighth of an inch makes the steel work difficult to bring together, it behooves the builder to see that the field engineering is well and truly done.

About the time that I was given full charge, the foundation work had fallen behind schedule. It was a night and day job to put on the finishing touches to make ready for the steel, then on cars rolling inexorably toward the job, and we knew that steel had to be unloaded without delay when it did arrive. The erecting gang had been organized, but a riveting foreman would be in demand a few days after the derricks left the basement. It was a deep, narrow hole, with an old-fashioned brick building along the long side, and little opportunity to guy a derrick. I was a cub, full of energy and inexperience, but skyscraper building leaves no time for deep meditations, and so, when a long, lanky Englishman tapped me on the shoulder, as I was overseeing the derrick erection, I did not hesitate to parley with him and in a few minutes I had a riveting foreman. It was Sam Parks, that debonair Robin Hood of the building industry, who was to rise to command of the New York building trades and die in Sing Sing prison. Sam produced a pair of overalls, and within an hour his bellowing voice resounded in that deep excavation, and I knew that I had a leader.

Steel was due the next morning, and the first pieces that had to be set were some fifteen-ton girders to be embedded in the foundation as a starting point for two rows of columns. These girders spanned the narrow dimension of the lot, wall to wall, and the deep basement made the problem of lowering them into place a difficult one. Just when the last touches were being put on the derrick, it toppled and fell, due to a faulty lashing, and to cap the climax, the erection foreman was slightly injured by the crash, and the mast of the derrick broken. There we were, hamstrung, and the cars bearing those girders already in the freight yard. It was Sam who came to the rescue.

All that night we worked, clearing up the wreck and preparing for the oncoming steel; but it was impossible to get a new mast. Sam asked what was in the upper part of that old brick building adjoining, and together we investigated by burglarizing the back door at three in the morning. After appeasing the watchman we prowled around with lanterns, and found that the third floor was a partly vacant loft. Sam went back and brought a couple of men with crowbars, and we put them to work digging out the bricks, so that in a few minutes we had a hole looking out into our excavation through which a man could put his head. Then we took the topping-lift off the broken mast, and with lines passed out through the hole, we raised it up preparatory to improvising a mast out of that old building. As the sun was rising, we were hauling timber up into that loft to strut and brace it for the strain of its old life. The store in the ground floor opened up, and the manager came running to see what all the pounding and commotion were about. I fear that I aided and abetted Sam as he swiftly invented a whimsical tale and waxed eloquent about how the falling derrick of last evening had undermined the foundations of the old building. The sure way to save it, said Sam, was to truss up the third-floor loft. The manager looked a bit unconvinced; it didn't sound just right, but somehow we mollified him. We put timbers on the front of the building and carried cable lashings from them to our improvised topping-lift. The store management, becoming incredulous, thought they ought to have their own engineers look it over. We agreed, but urged that they come the next morning. Through the day we improvised a foot-block, and by nightfall our makeshift derrick swung easily from the old building in anticipation of a good night's work. Sam's question to me was, "What'll she stand?" "Lord," I said, "if I ever figure it, we just can't do it. She won't stand up to the arithmetic. I've ordered the girders here for midnight, so we can have the side street to ourselves and block traffic while we're unloading."

Talk about great moments! Just at midnight the first truck rolled up. Sam stood on the bridge and signalled the boom out. The great hook on the fall hung over the middle of the girder with the lashings all on it. Then the heavy ring in the lashing was put over the hook, and there Sam stood in the glare of the floodlights, signalling the hoisting engineer. Traffic stopped breathlessly; the falls tightened; then the lashings creaked, and everything was taut. It seemed ages, and the men fell into silence. Creak, creak again from the lashings; bits of plaster from the old wall rattled on some planking deep down in the hole. I had posted myself at the timbers on the outside of the old wall to observe what the gruelling strain would do, and to signal Sam if the wall showed signs of caving in. But like a Viking he stood, complete master of that daring escapade. The great girder gently lifted off the truck, the men on the tag lines strained, as the massive piece swung slowly through the air and paused over its position. Then, with the same mastery of the situation, Sam gave the signal for lowering, and the engineer, standing alert on the levers of the brakes, let the drum slowly unwind; and the great girder settled into place. As the strain on the falls eased, cheers went up from the men who had stood silent, watching. Sam was calm, but triumphant. In that moment he had won the leadership that carried him into one of the most picturesque careers that ever befell a labor leader.

Parks was an expert rigger, which suggests that he probably had been a sailor, but I know nothing of his antecedents. In 1899, the building trades in New York were loosely unionized. The bricklayers were strongest, but even they were not aggressive. Such fight as the men showed was in behalf of a principle, the right of collective bargaining. Their leaders were talkers rather than doers, labor pedants eloquent of economic theories of unearned increments and wages. The employers still were top dog and, in general, intolerant and more or less contemptuous.

Sam was burdened with no theories; he was a doer, fighting as much for the love of fighting and of power as for anything. First rallying about him his devil-may-care iron setters, he seized the loose and flabby Building Trades Association and forged it into the powerful, militant organization it has remained ever since. A born guerilla, he fought the employers, not frontally, but by sudden forays. He introduced into the sympathetic strike sniping and like Fabian tactics, whereby a strike was called on one job at a time, permitting the unions to support the strikers easily out of the earnings of the men still at work.

He rose finally to be head of the whole Building Trades

Diagrammatic picture showing the underpinning of the National City Bank, New York, to permit the construction of the William Street subway.

Looking down an open caisson—for the Book Building, Detroit, Mich. Open caissons were sunk to a depth of 130 feet.

Pretesting a 1200-ton footing. This spread footing in a 26-story building, designed to carry 1200 tons, settled more than six inches. To prevent further settlement it was pretested. (See footnote on following page.)

Courtesy of Spencer, White & Prentis, Inc.

Council, and, reckless in this power, precipitated a great general action, the seven months' strike of 1903, which ruined many builders and, indirectly, carried Parks down with them. He had been doing a brisk private business in selling strike immunity to employers, and he was convicted of extortion and sent to Sing Sing, where he died of tuberculosis. Publicly, labor proclaimed Parks a victim of a capitalist conspiracy. Privately, I think, it admitted his sin, but forgave him his personal exploitation for what he had won for it, for labor held and expanded upon the gains that his militancy had brought it.

Two seventy-two inch plate girders were rivetted to the column and beneath these were placed four pairs of 15″ I beams. Twenty one-hundred ton hydraulic jacks were set between these grillage beams and the footing and cross connected to an electrically operated hydraulic accumulator and pump, and also to hand-operated pumps. By means of these, a pressure of eighteen hundred tons was exerted on the footing, depressing it several inches and compressing the ground beneath it. Pressure was maintained constantly on the jacks until forty-eight hours had elapsed without further settlement of the footing. The column was then wedged against the footing.

Courtesy of Starrett & VanVleck, Architects.

Eight West 40th Street. New York. Probably the first to use the pointed roof to enclose elevator machinery, tanks, etc., an accepted criterion of the best solution of the "inside-lot" problem. (See text, page 101.)

CHAPTER VII

CONTRACT RELATIONS

SKYSCRAPERS are built under two forms of contract, general and divided. The latter is a survival of simpler times when simpler buildings were parcelled out to independent sub-contractors, and the co-ordination left to the architect or engineer, even the owner sometimes assuming the builder's rôle.

The general contract is the usual form used on large metropolitan structures, and the one which best illustrates the work of a modern building organization. Under it an owner turns over the plans and specifications for a building to a single agency, and that agency binds itself to deliver within a time limit a completed structure ready for the tenants to move in. The contractor finances the work from month to month, the owner, however, paying the contractor a proportion of the actual outlay as the work progresses. The contractor buys and assembles materials, lets the sub-contracts himself, may himself perform certain of them, such as foundations, masonry, structural steel, and carpentry, supervises and administers the whole, and protects the owner against all contingencies, except the contingency of the owner changing his mind.

This can be more clearly explained by quoting from a paper recently read by Ward P. Christie before the American Society of Civil Engineers.

"Contrary to popular conception," Mr. Christie said, "the principal function of the general contractor is not to erect steel, brick or concrete, but to provide a skilful, centralized management for co-ordinating the various trades, timing their installations and synchronizing their work according to a predetermined plan, a highly specialized function the success of which depends on the personal skill and direction of capable executives.

"The mere broker cannot do this successfully. If one branch of the work falls down, he cannot perform it himself, but must seek a new agency which, at best, means serious delay. In some instances engineers and architects have successfully performed this management function, although they have done so, not by reason of their technical skill and training, but in spite of it. They were good construction executives, as well as technicians, with an understanding of business management—a combination rarely found either in business or the professions. Ordinarily the professional engineer's or architect's management of construction may reasonably be expected to succeed about as often as the contractor's execution of engineering or architectural design "

Unless the owner, architect, or engineer happens to be such an unlikely combination, the attempt to build a large structure on a divided contract is equivalent to trying to operate an army without a staff. Building is unlike any other form of industry, in that, like an army, it is a field operation, never under a roof, with many long and exposed lines of communication, and made up of a complexity of specialized units. It is a staff operation, a problem in logistics, as soldiers say—the getting up of supplies, the performance of a task and the removal of the waste.

Sub-contractors on a large metropolitan structure, operating each at his own convenience, bring about chaos. The work of each interlocks with that of the other; the thing cannot be done except in the proper order, and where no order is enforced, confusion and chaos ensue. Some jobs have been muddled through in this fashion, but the cost in time and money has been enormous, as some bankrupt owners can testify. For example, conduits must be placed when forms and structural parts are ready to receive them, not when the electrician feels like doing it. The heating plant is essential to the drying of the building where thousands of tons of water have played their useful part in the masonry of every sort, but which must be evaporated before the final finishing is done and, of course, before the building can be comfortably occupied. Sidewalk beams cannot be set until certain special ventilating equipment is in place. Proper safety measures are impossible under divided responsibility. The sub-contractors, put to extra time and expense by interference and the injury of workmen, can recover from the owner where no financially responsible agency such as the general contractor stands between him and them.

Under a general contract, if either time or cost exceeds the contract figure, the contractor bears the loss. Under a divided contract, where the architect oversees the whole, he is paid for no such responsibility and accepts none, and the loss is the owner's. The exception to the above is that many builders will not bid competitively for an operation nor take it at a pre-arranged price, believing that such bidding brings their interest and that of the owner into direct conflict. Instead, they take a contract at cost plus a management fee. Under the guidance of a skilful builder, co-operating with an intelligent architect, this system reaps rich reward to the owner in economies impossible under any other plan.

There is much misunderstanding as to just wherein economies and profits of the building industry lie. The feeling in the public mind that the builder makes inordinate profits by the shrewdness with which he conducts the building operation at the site is erroneous. The large economies of construction are made when the plans are being drawn, and the well-informed owner who avails himself of the valuable advice that may be obtained from the co-operation of an intelligent architect and skilful builder is really the one who profits in building operations. Such profits are not as spectacular as those which seem to come from an owner's being able to contract with the builder for the fulfillment of his finished plans and specifications at a price somewhat less than he had anticipated. He may chuckle with glee to think that he has bought his building for fifty or even a hundred thousand dollars less than he expected it would cost; but bound up in that set of plans and specifications, in all probability, is another hundred thousand dollars of losses that the owner must bear, due to the rigidity of his scheme and his failure to consult a capable builder while those plans were being drawn. If this latter course had been followed, the savings apparent in his competitive bidding, if they were legitimate, would automatically have come to him, and with this would have come the other economies that the builder could have introduced—now forever lost. This process has nothing spectacular about it, for when the owner, builder, and architect have, through coöperation and study, agreed upon all of the details of construction and these have been incorporated in the plans, the budget is pretty nearly the minimum. The owner may derive some further satisfaction by some very close buying on final closings of sub-contracts, and will surely reap the benefits that further conferences will bring as the operation proceeds, but these are small economies; the big economies were made before the owner's fortune was jeopardized.

Morton C. Tuttle, of Boston, one of the best-posted builders in the country, writing in the *Architectural Forum,* says:

If cost is a vital concern, and if it must be controlled, then, obviously, someone fully competent in cost matters should be constantly in contact with every step in the development of any design whose fulfillment in structural form involves cost . . And the more the nature of the design is such as to appeal primarily to the designer's creative imagination, just so much the more are considerations of cost likely to be overlooked. . . .

Speaking of estimates, he says:

Such an anlayzed estimate makes it possible for the owner to weigh the importance of each feature in terms of its cost, and to retain this feature or to discard that, according to his judgment as to its relative dispensability. The plan thus generally outlined, estimated, and finally approved serves to establish a budget by which the detailed development of the design can be controlled . . . Furthermore, in the end, a building plan developed in conformity with a preliminary budget, for whose application the cost expert is accountable, is pretty certain to present no necessity for those hasty and disfiguring last-minute modifications and eliminations which harrow the souls of all such designers as take worthy pride in their work. . . .

And further on, alluding to the co-operation of builder and architect:

Under such circumstances it should be understood that he is to work *with* the architect,—concerning himself, first, with matters of cost, as they develop in the design proper, next, with purchasing; and lastly, with the conduct of construction. Thus engaged, the contractor will often be found to serve as a reliable cost expert, and in any event as so efficient a supervisor of costs that through his agency the owner may from first to last feel confident of controlling his expenditures. So long as the general contractor is not subjected to the pressure of competition, he can work in this way; but, as soon as competition is forced upon him, his interest shifts, and such knowledge as he possesses is reserved for his own protection, not for that of his employer. . . .

If we must have a competitive contract, the logic must be faced; he says:

It may, of course, be argued at this point that the architect's specifications, by which the contractor and sub-contractor alike are bound, will stipulate the exact character of the work to be performed and the quality to be achieved; and that careful inspection should suffice to insure adequate fulfillment of these specifications Any one possessing a sense of humor might delight in following this optimistic theory in its application to a surgical operation, to the painting of a portrait, the composing of a piece of music, or even to the humble yet subtly exacting process of making an apple pie. In his inner consciousness, every intelligent person is aware that the one chance of obtaining good work is to entrust a task to the competent. Inspection offers no substitute for honesty and ability.

There is a general misconception in the public mind that builders grow fat on extras and changes. The very reverse is

true. Extras and changes are the bane of the building industry, and a large part of the wear and tear of building arises out
of the failure on the part of the owner to understand the builder's problem. He has accomplished so much in the way of
wonders, it seems incredible to the owner that he cannot everlastingly work miracles. To stop a gang of high-priced workmen when they are about to undertake a task in order to
change some small and seemingly insignificant part of that
task, and then judge the apparently large expense by the insignificant item changed, is where misunderstandings start.
The owner is told that it will cost $200 to move a doorway,
and he immediately protests that he is moving it only six
inches, seeming to imply that a $200 price ought to entitle
him to move it at least several feet. In his over-wrought state
of mind, it is almost impossible to explain to him that it would
be cheaper to move it a dozen feet and get away from a lot
of electrical conduits, plumbing pipes and other facilities, than
the mere six inches that he is now requiring Such a partition
change has been known to involve all of the following trades:
the mason, the plasterer, the sheet metal worker, the plumber,
the tile setter, the marble worker, the ornamental iron worker, the carpenter, and the painter. Figure the skilled mechanics involved in this, each one waiting on another for this six-
inch shift in the location, and the reader can see why changes
cost so much.

Eighty per cent of the cost of the usual skyscraper lies in
what is called the buy-out—the sub-contracts such as steel, elevators, plumbing, electrical, heating, ventilating, plastering,
painting and decorating, etc. Another ten per cent goes for
commodity materials such as sand, cement, brick and similar
items. Assuming the plans and specifications to be very accurate and complete, among any group of four or five competent builders, the bids on this ninety per cent of the work
will be nearly identical, unless some of them gamble on their

profit by selling short on futures. The remaining ten per cent of the building cost is the builder's direct payroll, which usually includes the foundations, masonry, bricklaying, carpentry, etc. He might perform this fraction carelessly and make it cost a little more, or he might perform it so skilfully as to cut the cost of this payroll work as much as ten per cent. The difference either way would be only one per cent of the whole cost. The price of a structure having been estimated scientifically, the owner's interest should be not how cheaply, but how well his agents build for him. With a skilful builder, an owner may feel confident that the economies of this direct work will be obtained—that goes without saying. But the tremendous advantage an owner gets comes from the intimate knowledge the builder furnishes as to methods, markets, purchase, availability and delivery, for under this plan the greatest elasticity of decision is available, and moreover, the skilful builder is working for the owner, not for himself. The public seems slow to realize that builders are not vendors of buildings, but are expert managers and co-ordinators, as Mr. Christie has said.

There are three major divisions to the work. The first is in the architect's office, where the owner has outlined his requirements. These requirements are technically known as the program. It takes from three to six months to prepare a complete set of architectural drawings, and even these must be supplemented and amplified as the work of construction proceeds. Preliminary sketches will be made by a competent architect within two or three weeks, but they are not sufficient basis for a fixed-price contract, and can only serve to define the general scope of the work. The owner who demands competitive bidding and a fixed-price contract must await the completion of final plans and specifications if he would avoid the peril of an incomplete and inferior structure due to the *caveat emptor* of his bargain.

The second stage is the business and managerial phase, the buy-out. For six weeks or more the builder's office is attended by sub-contractors consulting the plans and preparing their bids. These items may cost either more or less than estimated, but always they are the major expense of building.

In the third stage, the work is marshalled in lockstep and performance is the test, with the wrecker leading the procession of sub-contractors. Here the skill and generalship of the builder are shown. A thousand and one details are constantly before him. The sub-contracts, drawn with due regard to the rights of the owner and sub-contractor, nevertheless need administration. Certain work cannot start until certain other work has been completed. Conflicts of space and order of precedence are constantly under adjustment. The most obvious solution is not always the best, and frequently the job moves forward or lags through the mere attitude of the administration toward the "subs." The builder's experience and standing give force to his orders and decisions; his knowledge of the business and reputation for fair administration of his own work and that of the sub-contractors furnish the leadership and driving force that make for swift and sure progress. The capable builder is responsible for the whole progress of the work, and he stands between the sub-contractors and the sometimes capricious petulance of both the owner and the architect.

CHAPTER VIII

IMPORTANCE OF DESIGN

THERE are two ways of subordinating income to design in business building. The first is confined largely to banks. Not infrequently banks erect low classical structures solely for their own use on very valuable corners. Or they do the equivalent—erect office buildings, but reserve the lower floors and convert this expensive space into high-vaulted banking chambers. Banking can be done under a ten-foot ceiling as efficiently as beneath a sixty-foot ceiling, but in either case the banks have known what they were about. The public demands both an impressive façade and a marble interior of an institution where it deposits its money.

The second method carries a building to uneconomic height and perhaps spectacular elaboration as well. This may be a Woolworth Tower in New York or it may be, for example, a twelve-story office-building in a town where neither land values nor the demand for office space justifies such expense. In all cases, the motive is the same—prestige. Curiously, small-town skyscrapers frequently justify themselves economically, their prestige together with their more modern accommodations emptying older buildings of their tenants.

My argument is that the skyscraper is an intrinsically beautiful form. There have been beautiful and hideous skyscrapers and many that lie between, and the fault and the merit alike have been the architects', with rare exceptions. If a "prestige" building fails of beauty, the fault obviously is that of the architect, for the men who pay the bills are lavishly generous of confidence and money.

Striking beauty may be impossible to a purely commercial building; but if such a structure lacks poise and dignity, the fault again is that of the architect. He may be held by the owner to a stark simplicity of outline; but if the cost limitations can be met at all, they can be met with self-respect. The difference between grace and ugliness in a severely plain form is one only of a few thousand dollars, the architect's ability and a few lines in the right place. If an occasional owner should hold out for unrelieved common brick as a facing for his skyscraper, or insist on the exterior of a two-story Main Street shop for a thirty-story structure, a self-respecting architect can throw up his commission.

Amateur critics of skyscraper architecture have a mental picture of an owner holding the struggling architect's nose to a grindstone on which is carved an emblematic dollar mark. The normal owner builds for profits and he is unlikely to be a competent judge of design, outward or inward; but the true picture of his meeting with the architect usually runs like this:

The owner draws up his chair and his first words are: "I want something adequate and plain, no gingerbread, no fuss and feathers, mind you." It lies in his mind that excessive expense in building takes the form of visual ornamentation. It almost never does. A skyscraper is like a great passenger ship —the difference between a crack express liner and a secondary steamer is not one of hull and funnels, but of internal machinery and facilities.

The owner of a new skyscraper makes his choice between low and high initial expense when he comes to decide on elevators, plumbing, ventilating and heating equipment, internal facilities and finish. They may be good or bad, costly or cheap. When this point is reached he has undergone a psychological change. The visual appeal is so strongly ingrained in human nature that, having discovered by now that it is a relatively minor item in costs, the owner is willing to accept any amount of exterior design the architect wishes, may even urge him to more elaboration; but he has grown niggardly

about pumps, valves, copper, elevators, and the like. The real
optional expense is hidden inside the building and is neither
spectacular nor particularly beautiful. These things do not
stir his pride and imagination, and from a no-nonsense business man he has become a romantic, concerned with non-
essentials.

It was said a paragraph back that excessive expense almost
never takes the form of ornamentation. The word "almost" is in
recollection of one of the world's most magnificent banking
rooms, which began with the stern intention of being as plain
as a pipe stem. This bank illustrates not only an exception to
the rule but the shift in view-point of an owner. The bank
had grown wealthy and powerful over many years in an off-
corner of New York. A new financial district springing up,
the directors decided to shift the main banking offices to this
district. Architects were called in and commissioned to design a building.

"We are plain people down here," the directors said, "and
moving up-town will not change our natures. We want a simple, honest office-building that will earn its keep, and on the
main floor we want as simple a banking room as you can design."

"Right," agreed the architects, an able, imaginative and a
far from simple firm. "In order to get the utmost of simplicity
we will agree on the dimensions of the banking room and
leave it just four brick walls for the present." The directors
applauded this unexpected sanity. Thereupon, step by step,
the architects began to educate the board in the perfections of
interior design. As the directors succumbed to the beauties of
one suggestion, the designers pressed another. After six
months the board had agreed to one of the most lavish rooms
ever seen. It cost very close to $1,000,000, and by its splen-
dors forced a redesigning of the public entrances, elevators
and other details of the office-building to bring it more into

keeping with such a child. The final result was a monument
of which the directors are inordinately proud.

An individual owner naturally could seldom support such
magnificence, however he might be attracted to it; but inasmuch as a good architect, by his arrangement of elevators and
corridors, can get as much as twenty per cent more return out
of a given building than can a poor architect, owner and architect usually can meet on the common ground of good economies and agree on design.

Too much, not too little, ornament was the prevalent of-
fense of the skyscraper from its birth down to the war. Feeling their way about in this new architectural form, designers
took their æsthetic yearnings out in cheap and easy over-elabo-
ration, thereby robbing the structures of their two inherent artistic integrities—simplicity and power.

Ornamentation always is a lesser and frequently a questionable form of beauty. It is ever apt to bear no relation to
the building, or to be overdone, as a woman with too much
of the wrong kind of jewelry. Only great architects may be
trusted with it.

We have to thank neither architect nor owner, but labor,
for a better day. It was economic necessity that removed the
lace and embroidery from these athletic figures. Ornamentation still is cheap, but labor has become expensive. When
both were cheap and the skyscraper was new, heavily chiselled
festoons of granite and marble garlands of fruit and flowers
hung over the windows and doorways of our business buildings. Friezes and cornices invited festoons and brackets. Car-
touches were epidemic after 1900, and any structure of pre-
tensions to elegance dangled with festoons of abundance and
cornucopias spilling out a ponderous plenty wherever a vacant
surface offered. Inside, the allegorical horn of plenty was pursued with cast-iron pomegranates, figs and garlands almost in
reach of the plucker, but forbidding and ominous in their

metallic solidity. Ceilings were coffered and subcoffered with
massive conventional floral motifs looking down between
more ponderous plaster beams, themselves tortured with endless running ornaments, and the whole topped off with gold
leaf—an architecture that has come to be known as Early
Pullman Car.

Such florid elegance sounds expensive, but was not so, relative to the total cost of the building. If, however, carved
granites and marbles were prohibitive, there was terra-cotta.
An ancient building material, terra-cotta was restricted to
appliqué ornament before the skyscraper, because it crushes
under considerable weights. The curtain wall of the skyscraper, supported at each floor by the steel skeleton, made it
structurally feasible. It came into use immediately, and in
Chicago in 1894 the Reliance Building employed it exclusively in the curtain walls for the first time. Lighter than
stone or brick, highly fire resistant and sturdy if not subjected
to heavy burdens, it also is very cheap and capable of infinite
ornamentation, coloration and imitation of more expensive
materials. Offered any amount of flamboyance at a price
cheaper than simplicity, architects and owners were tempted
often beyond their powers.

Here labor stepped upon the scene, all innocent of artistic
intentions, and enforced a classic moderation. Such gewgaws
were expensive in labor if not in material, and the pruning
hook began to trim them ruthlessly from the tree of architecture.

CHAPTER IX

DESIGN AFFECTED BY ZONING LAWS

Up to now, like the owner, we have been looking at the skyscraper from the street. There are other angles of vision—from a distance, from other buildings, from the air. From any of them, the commercial high building was caught at an embarrassing disadvantage. By economic necessity, ordinary owners were forced to crowd their structures close to the limits of legal permission. No wasted space meant a cube rising unbroken from sidewalk to a flat roof. However graceful a façade the designer might contrive to woo the eye from the street, the mass effect of such forms was monotonous.

But the single worst crudity of the skyscraper was this flat roof crowned with one of the world's most unlovely sights—a naked water-tank set on stilts. The owner did not see his roof, neither did his tenants; therefore it was one with alleys and back yards. Anyway, water-tanks always had been set on roofs since buildings had running water. Then, as the electric elevator began to displace the hydraulic types, a further eruption occurred on the roof—a penthouse to shelter the motors which, in the better installations, are set at the top of the shafts. These early penthouses had the grace of a woodshed. The first step toward reform on the roof was to combine water-tank and motors in a penthouse, and so soften the lines of the latter and blend it a little into the building.

Probably we are justified in giving the credit to Goldwin Starrett, now dead, for the true answer to water-tanks, penthouses, chimneys, steam vents, exposed pipes and all roof excrescences in the 8 West 40th Street building, New York,

100

where he hid them all in a symmetrical, peaked roof. There had been occasional peaked roofs in the high buildings long before that, but purely ornamentally intended. Incidentally, this building is generally regarded in the profession as the best solution ever made of the trying problem of an interior lot.

The New York zoning law, passed in 1916, was practically, not æsthetically, intended. Depending upon the width of the abutting streets, the law requires that a building be stepped back at certain heights. These restrictions apply to three-quarters of the ground area of any new building. On the remaining quarter, a tower may be carried to any altitude the owner may desire. The law was intended to protect the rights of lesser buildings and to permit the sunlight to reach the streets a greater part of the day. Its principal, though purely collateral, effect, however, was to give to architectural design in high buildings the greatest impetus it ever has known, and to produce a new and beautiful pyramidal sky-line

No longer can a florist's box stood on end be erected in New York except to very limited heights. Speculators and other purely commercial builders and lesser architects had form thrust upon them by law. Abler architects were not unprepared for such an opportunity and rose to it splendidly. The 27 West 43d Street building in 1917–18 first illustrated the setback. This New York-born architecture is an adaptation of no other; it is our own, expressing ourselves. It is the sounder for having a reasoned motive rather than individual fancy behind it. Beauty of line and form, rather than beauty of ornamentation, distinguishes it. With only ten years' history behind it, the setback is a thing of grace and sweep. There is no reason to suppose that it is a finished form; towers of unimagined beauty should rise in the 1930's, and that is far enough to carry any prophecy in these eventful years.

We borrowed the zoning law from Europe, where long ago it became customary to limit any commercial building to a height not exceeding the width of the abutting street, measured from building-line to building-line; and as the purpose was æsthetic, the limitation was absolute. The setback is an American compromise. It is in effect, with local variations, in more than half the American cities of 25,000 population or more.

Washington is our only city, to my knowledge, that follows the European practice of absolute limitation. Boston did so until 1928, when it adopted a zoning plan similar to New York's. Meanwhile it is a city of one tall building—the campanile-like Custom House, which towers over it as the Washington Monument dominates the national capital. About twenty years ago, the builders of the Westminster Hotel in Boston contested the then newly passed ordinance and carried their structure above the hundred foot limitation. The city sued and won, and the top floor actually was sliced away after it had been built. The hotel stands to-day without a cornice and with the roof-line running through the heads of the windows of what was next to the top floor.

The intrinsic beauty of the setback form has carried it to cities that put no limitations on building dimensions. The new Philadelphia Electric Building not only is stepped from choice rather than necessity, but multicolored flood lights make the setback tower a rainbow by night—a unique and striking experiment that is likely to be adopted elsewhere.

The workings of the New York zoning law are so complex that a new profession—that of consulting expert in zoning—has arisen there. These experts advise and interpret between the architect and builder, and the Building Department in the fashion of lawyers. As builders, it has led us to make isometric outline drawings of the legal possibilities of any given building site. We carry these drawings to the final setback and indicate the unlimited tower possible on one-quarter of the area. The owner then may see what form his structure would be

From a photograph by Chicago Aerial Survey Co. and Starrett Building Co.

One of the devices an architect uses to present his case. Upper picture is an airplane view of a section of Chicago; lower picture is the same with two large buildings, centre foreground, drawn in, this giving a complete idea not only of how the new improvements will look, but how they will fit into their surroundings.

likely to take at any given height and he does not mistake
setback necessities for design

The assault on the skyscraper in recent years has been on
practical rather than æsthetic grounds. Led by Henry H. Curran, a former president of the Borough of Manhattan, its
enemies charge it with creating outrageous traffic congestion,
unsettling land values and putting a disproportionate burden
on the municipality for fire protection, water supply and
sewage disposal. According to Mr. Curran, the skyscraper
is the villain of traffic congestion. According to Harvey Wiley
Corbett, a distinguished architect, it is almost wholly innocent.
Both are wrong, in my judgment. Certainly, a building housing 10,000 workers aggravates the traffic problem for blocks
around. But the high building is only one factor in a condition practically inescapable in modern urban life. The motor
car is a worse offender than the skyscraper, as is demonstrated
every day in such cities of relatively low sky-line as Los An-
geles. As well padlock Detroit. London and Paris both have
rigidly limited sky-lines and relatively fewer motor cars, yet
their traffic problem is similar.

We tolerate traffic tangles because we cannot help ourselves. Better traffic congestion than no traffic. The basic difficulty goes even beyond the fact that our streets were designed
for slow-paced, moderate, horse-drawn traffic. We forget that
pedestrians died daily under horses' hoofs in the traffic of the
80's; that horse cars were slow, cold, smelly, infrequent and
abominably crowded in the rush hours; that workers toiled up
as many as six floors to their desks; that mediæval and ancient
cities were swarming warrens. In other words, cities always
have been crowded and uncomfortable.

Our own peculiar dilemma arises from the fact that we
were born into a simple world, have been catapulted into a
highly complex one and have not yet adjusted ourselves to
such a mighty revolution. In the last century man began, with

an ingenuity he was not suspected of possessing, to devise machinery to lighten labor, annihilate distance, increase physical comfort and wealth. Every such invention tended to make the city a more advantageous, more comfortable, more interesting place in which to live and do business. But as fast as the city improved, fresh hordes poured in from town and country to take advantage of these social-mechanical benefits, thereby largely cancelling the gain. The process goes on undiminished.

Modern life and industry are organized on a basis of centralization. Machinery, of which the skyscraper is part, made this centralization possible, and New York and similar cities are its consequences, essential to its scheme. Every hour of the day and night New Yorkers and other city dwellers are outraged by physical discomforts which they might escape by moving afield. They remain, and instead, new population flocks in from less congested areas to share and add to these discomforts, testifying that the city offers more to the average family than it demands of them. Either we must accept the city pretty much as it is for the present or we must decentralize modern life, return to 1850—which is preposterous. Individuals here and there may revert to the simple life, the commuter may compromise with it, but society cannot.

For twenty years or more men have been drawing dream pictures of a many-decked city, one in which traffic would be segregated by kind and speed on various levels. The inspiration obviously was the existing tube, subway, surface and elevated divisions of rail traffic. These visions carried such an arrangement much farther to many-tiered streets and cross walks providing for motor and foot traffic. It still is a dream, and an impractical one in the light of this day's sun; but it is the best, almost the only vision of a traffic solution we have imagined to date, and it may be that we are seeing its beginnings in the New York Central Building, recently constructed.

Twenty-five years ago, while steam trains still operated into the old Grand Central Station through the original inadequate tunnel, two passenger trains collided in the tunnel with heavy loss of life. The disaster brought a smouldering quarrel between the city and the New York Central Railroad to a crisis, and resulted in an agreement whereby the road was to acquire large areas of valuable contiguous property, replace the narrow tunnel with a wide cut, electrify and build the present great terminal.

The expense of acquisition and construction was so great as to threaten the financial stability of that very prosperous railroad, and in casting about for relief, the novel plan of selling air rights to permit the building of commercial structures over the tracks in Park Avenue was conceived—oddly, by a firm of St. Paul, Minnesota, architects, Reed & Stem. The chief engineer of the railroad had come from St. Paul and called them in to advise him.

Under the skilful direction of the late Ira A. Place, general counsel for the terminal development, the plan was put into successful operation. Beginning with the Biltmore, a procession of great hotels, apartments, clubs and office buildings rose over the maze of two-level tracks leading into the terminal, and they return a revenue in air rights more than sufficient to pay the interest and sinking fund of the bonded indebtedness of the terminal.

The value of these air rights, moreover, appreciated sharply between the first and the last building erected. The Biltmore Hotel, pioneer of the development, pays only $100,000 a year to the terminal; the later Park Lane Apartments, $130,000; the Commodore Hotel, $175,000; and the Roosevelt Hotel, erected a few years later, $285,000.

All these buildings are unique in that they have no ground rights, existing by virtue of their leases to the air above the street level. Patently, such superimposed structures have no basements; they take their steam heat from the New York Central's power house, located at one end of the terminal improvement—which heat, by the way, is purely by-product, inasmuch as the power plant is a necessity to generate electricity for the terminal.

The track space was blasted out of the solid rock of Manhattan—a stupendous undertaking in itself—and as the tracks come in on two levels, it will be seen that the upper must run in part on a floor-like structure corresponding to the basement of the skyscrapers above. Engineers recognized that the vibration of the heavy trains would necessitate a special foundation for the superimposed buildings. Accordingly, the building foundations reach down through the entirely separate structure on which the tracks run, a sort of cage within a cage, with no points of contact, every column footing of each structure insulated from the others by ingenious construction which reduces to a minimum the transmission of the jar and impact of the trains to the footings of the skyscrapers.

A pedestrian on the sidewalk in the Grand Central zone may notice a curious separation of the base of the buildings from the walk. There seems to be and in fact is a slot of about one inch. If the pedestrian carries a cane he may confirm his eyes by thrusting the cane to its full length into the slot and be persuaded that the building is resting on air. He is simply observing the separated construction. The sidewalk and street rest on the railroad structure, the building on a structure of its own, running down through the former but unattached to it.

The New York Central's own building, recently completed, occupies the last available parcel of air-rights property in the development and brings to completion the plan so brilliantly imagined by these St. Paul architects. In this building vehicular traffic is carried through a great building above grade for the first time in history.

Park Avenue, one of the main arteries of Manhattan, is blocked between Forty-second and Forty-fifth streets by the terminal, and between Forty-fifth and Forty-sixth streets by this new office-building. Through-traffic northbound rises on a viaduct at Fortieth street, crosses over Forty-first and Forty-second streets, carries around the Grand Central Station on an upper-level drive, overpasses Forty-fifth street, enters the New York Central Building at the level of the second floor, curves and drops through it and emerges at grade at Forty-sixth street. Southbound traffic reverses this order. This great skyscraper stands at what appears to be the end of Manhattan's widest avenue, pouring a continuous stream of motor cars out of one portal and drawing them in at another. The nearest parallel that New York can show is the Municipal Building, through which Chambers Street passes, but at grade.

Chicago's new Union Station is being developed similarly as to air rights The Chicago station is not electrified, and a great smoke chamber underneath the Daily News Building and the Plaza, the first structures to utilize these air rights, collects the fumes and powerful fans carry them out over the roof. The new Union Terminal in Cleveland has the same general characteristic of superimposed skyscrapers with special problems of vibration and smoke. On Broad Street in Philadelphia the new thirty-seven-acre Reading Terminal warehouse building will straddle the Reading tracks, as the Elverson skyscraper, across the street, already does. The dream seems to take form.

CHAPTER X

TYPES OF BUILDERS

The large modern city has created a new phenomenon—
the speculative builder. All cities have them, but only in New
York and perhaps in Chicago do they deal regularly in skyscrapers. Speculative building is not the highest type of construction, neither is it all its name may insinuate. Its followers
are opportunists, but their buildings are sound in structure.
The vigilant Building Departments see to that. For that matter, any steel-frame building undergoes much greater strains
during construction than it ever will be put to in operation,
and the speculative builder has no thought of trying to erect
an unsafe structure anyway. What he may do to a greater or
lesser extent is to slight it in facilities.

These skimpings insure two things—a high return on the
investment and an excessive maintenance charge. The speculator is not interested in the latter. The demand for space is
such in New York that he expects to rent the building quickly; then, with an assured rental roll, he can refinance or sell
to an investor and get from under. The type of investor who
demands this abnormally high return on his money is either
too ignorant to distinguish between a shoddily and a soundly
finished structure, or, like the speculator, he is gambling with
his eyes open.

If pressed, the speculator may defend himself with some
plausibility. "What's the matter with it?" he will demand.
"It passes the building laws, doesn't it? The buyer will make
money on it, won't he? As for elevators and such, they are a

good deal like automobiles—obsolete in a few years. In twenty years the normal increase in the land value will have more than justified a complete refitting or even tearing it down to build higher. Why build for eternity when all experience indicates that twenty years or so is the measure of an ordinary building's usefulness in New York? What is that but waste?"

Or a speculative builder may turn a profit without turning a spadeful of earth. For example, he assembles a building site, putting up enough option money to hold the lots for a year, or he controls them on a contract of sale with a deposit of earnest money. Now he hires an architect to draw him an imposing picture of a skyscraper; if it is several stories higher than the Woolworth Tower, so much the better. This he sends to the newspapers with a statement that work will start next Thursday. The newspapers like pictures of high buildings, real or imagined, because we readers have a weakness for them.

The next morning a dozen brokers are waiting in the promoter's office to ask him if he wishes to sell. They have no purchasers in view, but they are confident of finding such. The speculator will sell, he states, if the offer is interesting. The brokers hurry out to work for him for nothing, the newspapers have advertised his property with publicity he could not buy. The site is a good one and the psychology attracts a buyer, frequently another speculator. The promoter takes his quick profit on the real estate and steps out. Such a building site has been known to change hands four times, each time at an advance, before it was built upon.

There are opportunities in New York, Chicago, or any other large metropolis, for an enterprising operator to run a shoestring into a fortune legitimately in one enterprise. Here is a hypothetical case, based, with some surmises, on an actual instance in which a man, with no investment whatever, constructed a great building and sold it at a profit of $2,000,000 or thereabouts to a large corporation:

A mutual-benefit society owned a valuable corner occupied by a small building used for its own purposes. The society was dissolved, its affairs involved, and the property, subject to a first mortgage of $500,000, its only asset. The members needed cash. Here the speculator stepped in with an offer of $1,000,000 for the property on a basis of $200,000 cash and a second mortgage for $800,000. With this contract of sale in his pocket, he crossed the street to an insurance company, convinced them that the property was easily worth $1,500,-000, borrowed $800,000 on first mortgage, paid the $200,-000 and the old first mortgage of $500,000 and deposited the $100,000 in his bank.

Now the speculator looked about for a tenant for a new building to be erected on the property. The Universal Utilities Co., a great corporation, required more office space and the lease on its present quarters would expire soon. The speculator was able to offer the directors a very desirable corner and a building of any height or design they might prefer. The details of the building being decided upon, the corporation signed a long-time lease, anywhere from twenty-one to ninety-nine years.

The speculator now capitalized this lease by carrying it to a bond house, where he borrowed all the money he needed to take up the life-insurance company's first mortgage and to erect the building with a surplusage. The bond house floated an issue of securities and distributed it among the public. The bonds were well secured and every one profited.

The thirty-story building was constructed on investors' money and the Universal Utilities Co. moved in. Now the speculator went to his tenant and said: "You are paying me $1,000,000 a year in rentals. I can show you how you can reduce this charge to $600,000 by buying me out. I am incorporated as the Fulton-Cedar Building Co. We have a $5,000,-000 bond issue that calls for $250,000 interest and $100,000 amortization yearly. As the amortization piles up, my company will become valuable. I will sell you my company for $2,000,000 and you can pick up the second mortgage for seventy-five cents on the dollar for cash."

The deal offered the Universal Utilities a yearly saving of $650,000, less maintenance charges, and the eventual ownership of the property at a cash outlay of $2,600,000,—the sum of the promoter's price and the discounted second mortgage, —and it agreed. The speculator withdrew with something like $2,000,000, for his enterprise, judgment and salesmanship.

Note that he did not make $2,000,000 out of building. It is a commentary on the industry that fortunes made out of it always are made indirectly. When an industry ranks among the first two or three in a great industrial nation and no one engaged in it makes much more than a living except indirectly, something is wrong. The answer is that building, while conducted with high technical efficiency, is, economically, the most disorganized major activity known to modern business, agriculture perhaps excepted. Building and farming linger in the economics of the nineteenth century, whence all but these have fled. It is as fiercely competitive as the jungle and it is at the mercy of the customer to an extraordinary extent.

The customer is a customer, usually, only once in his life. He rarely knows anything of building, and arm in arm with this ignorance walks suspicion. His intention is to keep his eyes open, buy as cheaply as he can and, having a contract, to depend upon the building laws, the architect's inspection, the ironclad plans and specifications, a surety bond for completion, and, not least, the fact that he has the money and the whip hand, to force the builder to deliver. Actually he invites the contractor to make what he can.

The low bidder in this dog-eat-dog competition has been forced to sell short on his subcontracts, in the hope of passing the risk on to the little fellows and thereby seeing himself home. For example, the plumbing bids range from $50,000 to $60,000. Inasmuch as the builder bid competitively, he takes the low bid, but with the mental reservation that he will beat the plumber down to $40,000. He calls the plumber in, conceals from him the fact that his bid is low and tells him that nothing more than $40,000 will get the contract. The plumber, caught in this vicious circle, cannot pass the loss on to his workmen, whose union protects them, nor to the material men, whose business is stable. If he needs the work desperately, he is tempted to accept the $40,000 figure and hope to worm his way through the specifications by interpretation and substitution.

If the builder succeeds in grinding down all or most of his subcontractors, he is safe and they are playing with bankruptcy. If he fails to unload on them, he is certain of a loss; and inasmuch as necessity forced him into such a position initially, his margin of capital is likely to be too meager to survive the loss. Such competition explains the fifty per cent of failures in cycles of five years in the building industry, as vouched for by the Associated General Contractors of America.

The economic waste under competitive bidding is terrific. It costs a well-organized, thoroughly competent building organization somewhere between $1,000 and $2,000 to prepare an estimate and analysis of a complete set of plans and specifications, and requires from two to three weeks of intensive work. Such a builder is in contact with twenty to fifty lines of subcontracting, with perhaps a dozen representatives in each line. A total cost of possibly $10,000 to $15,000 is incurred by these subcontractors in addition to the general contractor's estimating.

An owner who wishes to exhaust every possibility of the bidding market is apt to send for ten to fifteen general contractors, all presumed to go into the estimate in fullest detail.

It is probable that all these general contractors will circularize much the same market of subcontractors, and therefore the geometric progression of cost may not apply; but it is fair to argue that, on a building to cost between $2,000,000 and $3,000,000, the free estimating service alone furnished by the army of contestants arrayed by the owner will run from $25,000 to $35,000—all but a minor fraction of it dead loss.

An increasing number of architects and owners will not be parties to this wasteful demand upon the building industry, appreciating that the builder properly is a fiduciary agent as much as a first-rate architect; that things cost what they cost, and any endeavor to make them cost less carries with it the risk of making them cost more, while almost certainly jeopardizing the quality of the work.

To return to the Universal Utilities Co., why did such a corporation not oversee the building of its own home and save this high middleman's charge? Because it and corporations in general are unfamiliar with building and are timid about engaging in so foreign an enterprise. The directors have other and more important matters to think about and prefer to rent or buy on the market.

Because of this unfamiliarity with and timidity about building on the part of lessees and owners, professional builders sometimes propose and initiate new building enterprises, as speculative builders do. The distinction is that the speculator builds on his own, and usually builds first and finds tenants later. He is a speculator first and a builder incidentally. The professional first finds an individual or corporate owner or tenant, and then builds for him.

The usual picture of the genesis of a skyscraper runs something like this: "A" owns a corner vacant or occupied by old, outmoded, small buildings. After long consideration, he decides to put up a many-storied office-building there, arranges its financing, calls in an architect, states his desires, hires a builder and either engages a broker to find tenants or functions as the operating and renting agency himself.

It may happen, however, that a builder notes that the corner would lend itself to a specific improvement—hotel, office-building, bank, apartment house, theatre, department store. If he is a speculator, he attempts to buy an option. If he is a professional builder, he first looks about for a possible lessee for such an imagined structure, finds one who might consider it, seeks out the owner, suggests that a tenant might be found in advance, brings owner and prospect together, advises with them, takes the contract for construction and turns the completed property over to the owner within a set time at an agreed price or at cost plus his agreed fee.

There are three principal classes of lenders to whom owners and builders may turn. The cheapest money obtainable is the so-called savings-bank loan at five per cent or a shade under. Savings banks will lend up to fifty or sixty per cent of the value of a completed structure upon completion. When such a loan is arranged, it is customary to obtain the money for use during construction from banking institutions specializing in such business. The rate in New York fluctuates between six and seven per cent.

The insurance companies offer the next lowest rate of interest—five and a half per cent on completed structures. Some of them make what they term a building and permanent loan, advancing money for construction to responsible builders from time to time as the work progresses. The interest rate is higher to cover the cost of the necessary service of inspection. It also is possible on occasion to borrow money from estates by arrangement with the trustees. None, of course, lends up to the full value of the property or the whole cost of construction.

The bond houses dealing exclusively or largely in building issues are a recent and important development in the industry. They underwrite first mortgage bond issues at six to ten points off the principal sum, covering their profit, risk and distribution costs, and sell the bonds, bearing six per cent interest or thereabouts, to the public. Inasmuch as they turn over the principal, less their discount, to the trustee immediately, theirs is the risk of disposing of the bond issue to investors. The borrower in this instance pays all interest charges during construction, although he may not require a sizeable part of the sum for as much as nine months; under the first and second methods he pays interest only as he uses the money.

These bond houses have been widely criticized, as new institutions are apt to be. The criticism is not well founded, for they are a wholly legitimate and an invaluable addition to the machinery of building. They have tapped vast sources of new capital for construction, without which we could not have built on the scale we have done since the war. The fact that one such house has failed under questionable circumstances has no more pertinence than the occasional failure of a bank has to the fundamental soundness of banking.

The bond houses were brought into being by the inertia of the old-time lending companies, which, in the earlier days of big building, continued to subscribe to the fetish, inherited from England, that a lender never must seek out a borrower or encourage him, but should wait on the latter's supplication. The borrowing and lending of money in modern business are as mutual accommodations as the buying and selling of merchandise, but the ancient superstition that the lender granted a favor and the borrower received one persisted.

It was not long after the start of the intensive metropolitan development ushered in by the skyscraper that these great lending companies were allocating as much as $100,000,000 annually, which, by self-interest, they should have lent on real estate in order to diversify their assets. Consider the hypothetical case of a Chicago real estate operator of the period 1900–10. He had a $2,000,000 project. The controller of a lending company or vice-president in charge of investment knew of old that the operator was a legitimate, successful builder. He read in the newspapers of the proposed building. Across his desk daily came a mass of information on real estate and building conditions. There was little system to this intelligence, but it confirmed the lending official's surmise that the operator's site was a good one and the project soundly conceived. It was to the lending company's interest to find such outlets for its funds, but it made no move.

In due time, the operator appeared, hat in hand, laid his estimates before the lending company and asked for a sixty per cent, or $1,200,000 loan—a proper figure. The official glanced at the estimates and the first thing he said was that it was the wrong time to build. Regardless of whether building was up or down, the operator was certain to hear that the city was overbuilt and the business outlook gloomy. He left, returned in a few days and heard this story again. When he returned a third time, the attack had shifted. He was told that his building was overestimated. Perhaps the company could lend on a valuation of $1,600,000, perhaps not. Next the lender trimmed off this and shaved off that, reducing the deal to a David Harum horse trade.

The operator knew in advance exactly how this farce was to be conducted. If he was a bit unscrupulous, he estimated his building at $2,500,000, instead of $2,000,000, and after three or four weeks of dickering, the least he would take and the most the company would lend teetered into balance, all without too nice regard for the equities or for the actual earning power of the building. True, the tradition was that operators were apt to be slippery customers or vague-minded optimists upon whom it was wise to keep a tight rein, and the tradition had a basis in fact. As building evolved from a trade into an industry, however, the lending companies remained static.

The operator, under these circumstances, may have been driven to the then much higher rate of the early bond houses. As this new form of underwriting and brokerage became more stabilized, the rates were lowered, though they continue, naturally, to be higher than bank or insurance rates. When an operator goes to a bond house to-day, he is greeted with: "Fine! We want to do business with you if we can. That is a good corner and it looks like a good time to build."

When the modern operator lays his calculations of expected rentals on the table, the bond house refers to its codified charts and quickly verifies the figures, and checks them promptly against the recorded costs of fifty analogous buildings.

It is not here necessary to be hypothetical; let us cite a famous instance of how a speculative builder's vision shamed the unanimous judgment of experts who kept their noses too close to the copy-book of tradition.

In the course of building a subway, New York found it necessary to buy a block in the heart of the city. The subway curved sharply at a depth of fifty feet under part of the property and there a great excavation was made. When the subway was finished, the city offered the land for sale subject to the easement of its improvement fifty feet below. Now "easement" was a terrifying word in those days. Immemorially it had implied a bar sinister on the escutcheon of title. Lending companies averted their faces, and the property, so advantageously situated in every respect, lay idle.

Among the buyers who had considered its purchase was a concern which planned a new building in this neighborhood. The site was ideal for its purposes—the best available in all the region, but the directors choked on the easement. At length a bold speculator braved this sinister thing lurking below ground and bought the whole property from the city for $2,900,000. Thereupon he turned around and sold the less desirable interior half, which was not encumbered by a subway easement, to the very concern which previously had flirted with the property and skittered away, and they probably paid him considerably more than half the $2,900,000 he had given the city for it all.

While the concern prepared to improve its half with a high building, the speculator promoted another large building on his corner half. As is the case with promoters, he endeavored to borrow every cent possible by mortgage and otherwise to carry out his plan.

The usual sources of such money told him that he was offering damaged goods and waved him away summarily, forcing him to turn to an early bond house, where he borrowed $6,-000,000 and paid for it through the nose, the net proceeds to him being little more than $5,000,000.

These facts were widely known in New York and there was a great clucking of tongues. The enterprise was denounced in genesis, financing and execution as a reckless piece of speculative folly. A survey was cited, as one argument, to show that there already were several million feet of unrented space in that section of the city. Nevertheless, the building was eighty-five per cent rented before completion and was highly profitable in operation from the outset. Soon after it was opened, it was wholly occupied and offices rarely have been available in it since.

The next chapter was even more ironic. Within a few years after completion, the promoter, finding burdensome the drastic annual sinking fund which the original bond issue had imposed upon him, went to one of the great concerns of the country, just such as had balked at the easement and deplored the whole venture as unwise and unsafe, and borrowed on the property at a low rate of interest, $6,000,000, with which he retired the bond issue and paid the premium on the bonds which such retirement always involves. Few of us can hope for such a handsome retraction.

Since then he has been offered and has refused $11,000,-
000 for the property. There was nothing accidental in this increased value; no sudden shift in business channels or other
unexpected turn of events explains it. All of it was predictable
in the light of New York's previous history. The experts simply had sold short on New York and had been frightened by
a name.

Fiduciary concerns are the trustees of vast sums of others'
money and they are rightfully conservative. As the custodi-
ans of these great reservoirs of capital, however, they have a
secondary duty to industry. If they are to err on either side,
it is much to the public interest that it be toward caution, but
it is most to the public interest that they strike an intelligent
mean. Conservatism always is in danger of atrophying to re-
action.

The survey referred to was made by a firm of unques-
tioned integrity but of literal turn of mind—as might be expected of statisticians. They began with arbitrary boundaries,
taking in a district geographically compact but widely diver-
gent in land values and transportation facilities. Then they
painstakingly charted every foot of vacant floor space within
those boundaries and struck a total. A casual check showed
that of this total of 2,000,000 feet of unrented space, some
three-quarters was in old, even ancient structures, many of
them inconveniently located, and that an inconsiderable part
was in modern office-buildings.

Properly located modern office-buildings will fill quickly
in defiance of surveys so long as their competitors are old, outmoded and badly situated. The public seeks new, modern,
convenient office space just as it seeks the new, the modern
and the convenient in homes, automobiles, clothes and all its
wants. Fine new office space begets more fine new office space,
and the suffering falls on the old and outmoded. Skyscrapers
tend to grow obsolescent as do motor cars and skirts. The

problem of the owner, architect and builder is so to design and construct them that progressive maintenance will largely overcome the obsolescence, for, unlike motor cars and skirts, they have permanent, stabilizing advantages of favorable location.

Courtesy of Starrett Brothers, Inc.

Photograph of an accurately detailed scale model, six feet high, of the Wall and Hanover Street Building, by which the architects studied details, proportion, color of material, etc. Delano & Aldrich, Architects.

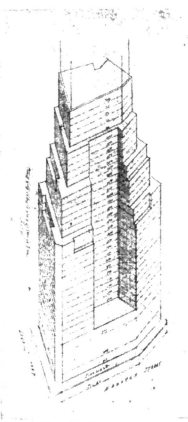

Courtesy of Yasuo Matsui.

Isometric study of Wall and Hanover Street Building, illustrating possibilities of setback under New York building law without regard to design or fenestration. See pages 71 and 102.

CHAPTER XI

THE GOOD OLD DAYS, AND DEMOLITIONS OLD AND NEW

MODERN builders have frequent occasion to smile wryly at the phrase, "Good old-time building." "Those were the days," sighs the romantic layman, and perhaps points to some beautiful piece of mahogany, let us say a fine old door, superb in its antique coloring, rich in its well-kept finish, but solid. "Solid mahogany doors," the sentimentalist says, and points with pride. The builder also admires its venerable finish and marvels at the painstaking labor that some devout craftsman of old put into it; worked out by hand with infinite patience and appreciation of its graining and color. The panels were made and finished separately before the door was assembled. The mouldings were worked out by hand with old-fashioned moulding planes, limited in their scope by a few standard and well-accepted quirks and ogees; perhaps the assortment of some honest Scotch tool-maker. But with all this craftsmanship, how was it put together? The stiles and rails were solid. "Of course!" says the admirist. Yet nothing could be worse for the door, according to modern standards. Our finest modern doors are built up with veneers, also selected for their beauty of figure and color, but selected from an infinitely greater stock and assortment than the old-fashioned joiner could possibly have inspected in a lifetime of labor. For modern lumbering selects and assorts, not from a few mahogany logs brought from far away India by some good trading clipper ship, but from vast stocks of imported, culled, selected and scientifically kiln-dried mahogany, where the choicest pieces from thousands of logs came before the discerning eye of a trained specialist in hardwood.

But to go on. The old door was "mortised and tenoned," and carefully clamped after gluing with the best glue available to the conscientious craftsman of old, but it was at best an apothecary's concoction of uncertain ingredients, with much hocus-pocus about its mysterious composition,—in reality a crudely prepared reduction of horse hoofs and gum arabic. So the old door was glued and pegged and varnished with a varnish about as uncertainly prepared as the glue. It graced a mansion and was the pride of the lord of the manor; but it started to fall apart about a decade after it was made. And when steam heating was installed in the manor house, the fate of the glorious piece of craftsmanship was sealed. It remains to-day the admiration of its owner because it has been glued and reglued, and revarnished, and indeed, had such lavish care bestowed upon it that veneration commences to take the place of critical inspection. The open joints in the old door add to its picturesqueness; its sag is unobserved or rejoiced in, and its oft-repaired hardware is just what it ought to be, although it must be handled gingerly.

The fact is that the old door is judged by a wholly different standard than its modern counterpart would be. Modern doors, built up on cores as we have seen, with beautifully selected veneers, with lock-jointed corners, with dovetailed splines between the alternated grains of wood in the cores, glued with scientifically prepared glue that is mixed to formula after painstaking research, are as much better than the grand old darlings as electric lights are better than sperm lamps.

Consider the good old masonry before the days of Portland cement. Again we have perfection of craftsmanship in the laying, but look closely. It was laid in lime, the best mortar then available, but cracked and crazed it is, handsome in its imperfections and venerable through sentimental association, but constructionally inferior. It simply would not pass the strength tests of any modern building department, although many cities permit lime mortar in the curtain walls of skyscrapers, recognizing that the structural steel fulfills the complete structural requirement and the walls are simply devices for keeping out the weather. In fact, all sentimentality aside, its cracks and imperfections in the shoddiest modern work would horrify the very sentimentalist who longs for the good old days.

When the wrecking begins, two or three men will drive crowbars into the lime-mortared joints of the massive walls of granite and marble backed up with brick masonry, and will pry them apart with one heave. Lime mortar was the best binder known in the good old days. The thin curtain wall of an old twelve-story steel-frame building will give the wreckers infinitely more work, for cement has welded the bricks into a homogeneous mass. Bricks will shake out of lime mortar like peanuts from their shells, are easily cleaned and are sold at second-hand. Cemented brick walls are battered down, breaking across the bricks more often than at the joints, and are carried out to sea—great masonry chunks, to be dumped as so much dead waste. With the mere curtain walls down and the hollow tile partitions and floor arches battered out, the wreckers will be confronted with the real structure, a frame of steel which can be cut apart only with the modern ingenuity of the oxy-acetylene flame.

These old buildings were massive, with space wasted both in their ponderously thick walls and in poor arrangement. They were the best building in their time and they were amply sturdy, as their long and useful lives prove, but no such margins of safety and endurance were built into them as is done as a matter of course in their descendants, the skeleton steel structures of to-day.

Demolition of some existing building almost invariably precedes the erection of a skyscraper, and such work has come to be a vocation in itself. Wrecking and the sale of second-hand building material go hand in hand, and if the buildings to be demolished are old, they sometimes yield a surprising amount of salvage in the way of brick, windows, glass, stone, lumber, doors and trim, to say nothing of the junk metal from pipes, radiators, plumbing fixtures, etc. The wrecker has outlet for these materials with a clientele and a market far removed from the throngs who idly watch him work and wonder what is to become of the vast amount of rubbish that roars down the chutes to the waiting trucks. Dust and refuse, made soggy in the vain effort to abate the nuisance, fill the air with a disagreeable acrid odor that characteristically bespeaks a demolition job.

Go to remote parts of the city, in the tenement districts, where racial colonies huddle together in out-of-the-way sections, where thrifty foreigners are making their first struggles with property ownership, and there you will find these second-hand materials being put to good use. Sometimes these structures are grotesque and laughable; sometimes they are put together with considerable effort at design and good arrangement; but they are to building what the wearers of second-hand clothing are to the patrons of the new and fashionable shops. Yet in the aggregate, these buildings of second-hand material absorb a large proportion of the salvage from the demolitions we see.

More and more are we seeing steel and fire-proof buildings demolished to make way for greater ones, and these yield valuable structural steel, motors, fans, etc. Again, the best of this salvage finds its way into lesser structures, the steel to be used as lintels and wall-bearing beams; seldom a column, however, both because columns are generally designed for a certain very specific location and floor height, with definite connections and base, and also because old columns are cut with the oxy-acetylene torch, leaving ragged ends.

A few efforts have been made to salvage the skeleton frame for use at another site, but in these cases it has proven of doubtful economy. In the first place, the building to be built in the new location would have to be exactly the size and shape of the one demolished. Beyond this, the basements, story heights and ground levels or grades would have to be the same. Elevator and stair arrangement would be the same unless the new owner desired to make structural changes in the field as the frame is re-erected, a most costly and unsatisfactory procedure. Finally, the floor plans of the new and old buildings would have to be about alike, due to the column spacing and general fixity of requirement of the original structures. Add to this the fact that the old structure probably was torn down because of its obsolescence, and the new owner has little room for consideration of re-use of the old frame.

I have known of one or two of these attempts to save the steel frames of buildings, thought to be so standard in construction as to seem surely usuable at some other site. Such salvaging involved most careful demolition. As each piece was lowered, it was carefully marked and recorded. At the storage yard, great care had to be taken in the stacking, for not only was it necessary to keep a record of the position of each beam, girder and column, but the storage had to be so conducted that as the members were stacked they would lie in their proper order for re-setting. Thus, the steel skeleton must be stored commencing with the roof. The misplacing of a few girders or a length of column would involve endless confusion. Consider further that if the material lies in storage any length of time, the identifying marks rust off or become obscured by dirt, and are washed off by rain.

The steel forming the main auditorium of the old Madison Square Garden in New York, I am told, was stored in anticipation of re-erecting it as an auditorium in another city. In my opinion, it is still a doubtful experiment in economy. The trusses were of an old and obsolete design, flimsy and wasteful of space, and last but not least, difficult to erect. Perhaps the backers of this experiment will some day disclose to the world how it all worked out.

There are some rare cases where particularly beautiful edifices have been removed from one place and set up in another. The notable example I think of is the removal of the façade of St. Bartholomew's Church from 44th Street and Madison Avenue to Park Avenue and 50th Street, New York. Sentiment largely dictated, and it is doubtful whether any economy resulted from the effort.

So we need waste no tears from an economic point of view over the demolition of beautiful or useful structures to make way for others, bigger and better; for such is building and such it always has been.

Courtesy of Starrett Brothers.

An outside scaffold containing all the temporary hoists built on the Wall and Hanover Street Building, New York, to save time in construction.

CHAPTER XII

EXCAVATIONS, SHORING AND BRACING

EXCAVATION seems a prosaic thing—just taking out dirt. But consider it in connection with retaining walls holding busy thoroughfares and great and complicated shoring and underpinning operations, deep foundations, water, clay, muck and rock, piers with elaborate foundation structures, and it commences to take on an imposing importance. Planned with care, it must go on uninterruptedly, with every move like the moves on a chess board. Here certain column footings must be sunk, first located accurately by field engineers in the roar and turmoil of steam shovels and derrick excavation buckets, the structural engineer's plans having precisely located these column centres with respect to the building lines and their depth below grade. When these few advance footings are completed, they are used to receive the thrust of the long shores holding the streets and adjoining buildings, that other shores may be removed to permit the building of more footings. And so it goes—planning, scheming, daring, safety and progress, all must be merged, and the inexorable time schedule and budget are ever before the builder's eyes.

Water, whether from the surface, from springs or from the tides, is ever the enemy of foundation work, but the builder may conquer it if he be alert. In some locations in cities on the seaboard, the ground is saturated from the sea and in a measure the water level in the ground rises and falls with the tide, but it must be conquered nevertheless, and the devices of the resourceful builder are many. The power of water to dissolve and its swift erosive action on stable soils make necessary the vigilance that we see in the preparation for a foundation job. Derricks, engines, timbers, planking, pumps, and all manner

ourtesy of Raymond Concrete Pile Co.
The Raymond concrete pile ready for driving.

Courtesy of Raymond Concrete Pile Co.
A Raymond concrete pile partly driven, showing the driving mechanism and the steam trip-hammer.

Courtesy of Starrett Brothers, Inc.

DRIVING CONCRETE PILES

Note the casings at the left; also the finished capping, right centre, and the men preparing a cap, foreground. Note underpinning of adjoining building which was done preparatory to pile driving to relieve adjoining building from ground shock of the driving operation.

of tools and appliances are there on hand when the work starts, for here is no place for waiting. When erosion and dissolving soil once start, there is no turning them back; the best that can be hoped for is to stem them, and the occasional street cave-in or damaged adjoining walls, generally bespeak lack of forethought or, perhaps more accurately, lack of sound engineering knowledge. Accidents will happen, of course, for at best, it is a hazardous undertaking It was Marshall Ney, I think, who said, "To be defeated is no disgrace; but to be surprised, a disgrace." Some such figure of speech could well be taken as the slogan of the foundation engineer.

Excavating is not just taking out dirt, even in the complex environment of which we have just spoken, for excavation is inseparably linked up with the construction of the scientific foundation. Consider earth and rock and water and then the agglomeration of all three in a single deep foundation. Added to these difficulties, adjoining buildings are nearly always with us; sometimes just plain, little three-story buildings with rickety lime-mortar walls, sometimes greater and more complicated structures. Whatever they are, they nearly always have to be underpinned and shored. Underpinning means carrying footings of an existing structure to a greater depth, implying at the same time the bettering of the original footings; and always the work must be done with the least possible disturbance of the existing structure. Shoring, sometimes properly called shoring and bracing, is just what that dual term implies. It is the placing of heavy braces against a wall or any heavy structure to give security and support. When a building is demolished, the adjoining buildings are frequently deprived of a needed lateral support on which they were originally designed to depend. Thus, the removal of one or more of a continuous row of brick buildings sometimes leaves unstable adjoining walls, and these have to be braced, or shored, as builders say. Such old buildings were frequently constructed with

Sinking a caisson. A heavy weight is required to overcome the ground friction of the caisson as it sinks, also to counteract the upper pressure of the air within the caisson.

Riding down a steel tube in close quarters. A trip-hammer is hung from the derrick fall, and must be manœuvred to drive the tube accurately in place.

Busy pneumatic caisson job. Foundations for the Federal Reserve Bank Building, New York. Note how completely the derricks cover the job. Bucket being lowered through a caisson lock, lower right.

"party-walls"; that is, a single wall, generally twelve inches thick, serving the structures on either side of it, and the property line was an imaginary one passing through the centre of the wall. Flues for the fireplaces of both buildings generally honeycomb these old walls, but at best they are insecure and are troublesome factors in the new building. Party-walls have an ancient legal status, with much legal lore about the rights of the parties, and also much legal rubbish, generally invoked by the owner whose building remains to be dwarfed by the splendid new structure alongside.

Shoring of streets includes the struts, braces and spur braces that press against the breast pieces or "wales" running continuously along the sides of the lot where earth is to be held from slipping. Sheet piling, as its name implies, is that continuous sheet of pieces laid edge to edge and driven down a foot or two in advance of the excavation which it protects. In simple work, sheet piling is ordinary plank, generally with the cutting edge somewhat sharpened to bite into the earth and facilitate driving. In more complicated and deeper footings, the sheet piling is of steel, specially rolled for the purpose with interlocking edges.

Once in a while an adjoining building has deeper foundations than the one about to be built, and the builder's heart rejoices, for one of the perils and anxieties of his calling is here removed; because, in spite of all the advance in the science of foundations and the improved facilities that modern invention has contrived, foundation and underpinning work is hazardous at best, and the sigh of relief comes to every one concerned when the foundation is finished and the walls are up to grade.

If we have access to the adjoining basement, and in some cases the law gives that access, underpinning is relatively simplified. Here we cut holes through the wall to be supported at intervals and thrust heavy timbers through, supporting the inner end on cribbing laid on the cellar floor of the old building. A series of these properly supported inside and outside on cribbing with the contact on each timber securely wedged up to insure a bearing, together with a few supporting shores, and excavating can proceed under the old wall, which now is carried securely on a series of these timbers straddling the excavation. These timbers are called needles. When very heavy buildings are to be supported in this way, the needles are heavy steel girders. Needling is not a common occurrence where heavy adjoining buildings are encountered, the more common way being to underpin the walls in short sections or piers; or to introduce steel tubes, jacking them down one after another until the entire wall is caught up. Then the excavation proceeds. The driving of these tubes is not so easily explained at this point, and we will leave the subject here to be taken up when we are watching the driving of steel tube foundations.

Thus it is seen that for the skyscraper, foundations are the all-important thing; the first concern of the builder, the work that demands his greatest ingenuity and contains his greatest risk; yet out of it he gets his greatest triumph. When bed-rock is encountered, down into bed-rock we go, where depth of basements, one below another, demand quarrying on a stupendous scale, with the added difficulty of trussing up busy city streets and channelling along adjoining high buildings. Here the work-gangs sweat and toil behind chugging, hissing air-drills; then the "Look out!" yelled and taken up from corner to corner of the excavation, a silence for only a moment; men stand behind obstacles; a quick push on the igniter by the blasting foreman; a dull booming thud, smothered under a blanket of heavy woven cable,—a hundred tons of rock have been dislodged. Then the rattle and roar of the pneumatic drills again, and in a trice the great excavation swarms with workmen back at their tasks.

Let us stand on the curbstone and observe the busy activity where a great skyscraper is to be erected. The demolition has been completed and perhaps a steam shovel or two are slashing at an earth bank and nervously whirling to dump their loaded dippers into waiting trucks. Shorers are working at the sheet-piling along the street lines, and already a maze of timbers and struts proclaim the battle that is expected, to hold the streets secure until the permanent construction is completed. Underpinning of the adjoining structures is apt to have been about completed, for the far-sighted builder attends to that among the first things he does. If he has been able to get access to the basement of the old structures a month or two before demolition starts, the underpinning gangs have been as busy as beavers in that dark basement carrying the footings of the old adjoining structures to a safe depth, in anticipation of the excavation and disturbance that will entail on the commencement of the work of opening up the adjoining lot for its deep basements and adequate foundations.

That shoring of the streets—how haphazard it all seems, and what utter confusion in the scattered piles of timber and planking! From the curbstone it will surprise us to know that the wales or breast pieces of the sheet piling have been placed with engineering accuracy to allow for the building of the curb walls, which must be waterproofed from the outside perhaps; and in their placing, the plan of just how that wall is to be built, how the workmen are to conduct the operation, have all to be taken into account. Then there are the "shores," braces of heavy timbers, generally placed slantwise down into the lot, thrusting perhaps against a previously built row of piers that this plan of action calls for. The "spur braces" run at diagonal lines from points about ten feet down on the shores, spreading on the wales like the outspread fingers of a man's hand. These are to keep the wales from bowing inward under the terrific pressure that street earth will impose on the sheet piling.

"Confusion and seeming disorder," you may say. "No; order of the most deliberate kind," is the answer. The shores and spur braces have been located with a nicety to allow for columns and footings between them, and at such levels and such slants as will allow for the setting of columns and structural floor members. Often, before those braces can be removed, the pressure of the street has to be taken up by the sidewalk beams which, in turn, transfer the thrust through the ground floor and basement structural beams; and these must, perhaps, depend for further reinforcing upon the masonry floor arches. Thus these floors are to act as continuous plates to hold stable the pressure from the street, perhaps the ground water pressure as well. Through the sheeting and shoring, old Mother Nature has been tricked into holding still while the permanent structure is being built to hold her in place forever.

Courtesy of Starrett Brothers, Inc.

Kresge Department Store, Newark, N. J. Shoring and bracing and setting steel at one time. The thrust of the heavy shores is being transferred to the column footings and structural steel, which will allow the completion of the excavation along the line.

CHAPTER XIII

FOUNDATIONS OF VARIOUS SORTS

A PNEUMATIC caisson is a formidable sounding thing, and in fact bears implication of a subject too technical to be discussed here; yet it is used by the skyscraper builder to great effect in structures where foundations and excavations must be very deep, and ground-water, quicksand, and muck are not conquerable by any other means. Take an ordinary drinking glass and invert it, and then press it down in a bowl of water. It will be seen that the water level inside the glass is considerably below the level in the rest of the bowl. Here is a miniature pneumatic caisson. Now, if one could introduce a tube through the glass and sealed to it, and then blow on that tube, it would be observed that the level of the water under the glass would sink still lower. Here we have the pneumatic pressure feature of the caisson. The heavy, steam-driven air-compressor that puffs lugubriously all day long beside a caisson job is supplying the pressure that we observe when we blow on the tube. Now suppose we fill the bowl with sand, leaving ample water in the bowl also, and press the glass down. The level of the sand would remain the same both inside and outside of the glass, but, unseen, the water level in the saturated mixture would recede under the compression of the air in the glass. Add pressure from the tube, and the water level will be slowly forced down to the bottom rim of the inverted glass. And here we have the whole principle of the pneumatic caisson.

Courtesy of Starrett Bros. Inc.

Driving steel cylinders for the foundations of the L. S. Plaut Department Store, Newark, N. J. Four pile drivers, using steam hammers, were employed, and over 40,000 linear feet of cylinders were installed.

Courtesy of Starrett Building Co.

Not a gipsy camp but rain protection for a series of "Chicago wells" being carried down by the open caisson method, a development made possible by the soil conditions found in the Great Lakes region.

The great, air-tight boxes or steel drums we see in a caisson foundation job are only the glass enlarged, plus the air locks on top that allow men and buckets to pass through into the inside pressure, where the water is forced back by that pressure. The "sand hogs" dig out the muck and sand, load it into buckets and tap on the steel lock, a signal that all is ready to hoist. The bucket is pulled up by the cable that passes through a rubber gasket fitting closely around it and permitting the escape of the least possible amount of air. Up it passes through the first gate, which opens to let the bucket through. Then that gate is closed behind the bucket, thus impounding the pressure below where the men are working. The lower gate having closed, the upper gate opens and the bucket is hauled out into the open air and dumped into a hopper awaiting the trucks. Then the bucket is swung back over the lock, passes through the first gate and the gate closes behind it. Again the gasket clasps the cable, signal is rapped on the outside of the caisson, the men below rap back a signal of readiness, the lower gate is opened, and the bucket passes through into the caisson chamber, where the cycle of operation is repeated. Men pass in and out of the caisson much as the buckets do, and thus the air pressure inside the caisson is always maintained until the desired foundation is reached, generally either bed-rock or hard-pan.

Now this caisson is heavily loaded with a great superimposed concrete block, which is added to as the caisson sinks, and the caisson is allowed to sink only as fast as the sand hogs excavate ahead of it. It frequently happens that the weight of the concrete must be supplemented by tons of iron piled on top of it, great iron cubes which look strangely incongruous around a masonry foundation until their use is understood. The caisson has a sharp cutting edge of steel, and in sinking, it cuts its way just ahead of the excavators inside. The caissons thus sink slowly, but under a guiding skill that the builder furnishes. When the caisson finally arrives at the desired foundation, concrete is sent down in the same buckets to be packed in by the workmen inside. First it is packed against the outside edges, and then on the bottom, the workmen gradually receding toward the lock tube through which the buckets of concrete pass. The process is not unlike a man painting a floor. He paints from the outside, gradually receding to the door, from the threshold of which he puts on the last few brushfuls. So it is with the concrete packing of the air chamber. The superimposed block of concrete which furnished the weight to sink the cutting edge becomes the pier of the permanent foundation, and the concrete packed tight against it in what was formerly the air chamber forms a homogeneous whole, a solid mass of concrete from bed-rock to the desired height.

If it be the plan of operation to make a continuous wall around the excavation, these caissons are placed successively, one touching another like dominoes on edge. The joints between the caissons are sealed, and then the builder has practically unrestricted sway to dig out the material inside the great cofferdam thus formed, and the footings inside are built on bed-rock without further trouble from earth or water.

Wood piling has a very respectable antiquity, and if we follow the archæologist, we can connect it with the earliest forms of human habitation in the shell mounds of the lake dwellers, where the fragments of these submerged pile stilts are still to be found in the lakes of Switzerland. This is only mentioned to note the everlasting quality of wood when submerged and not exposed to destruction by organisms such as the teredo, or by water-borne wood solvents. Mediæval builders used them, it is recorded, in connection with heavy foundations, but their use could hardly be classed as any feature of a science, although a few great foundation engineers of the Renaissance no doubt gave scientific value to the uses they made of them.

In the early days of skyscraper building in Chicago, it was the practice to dig down a few feet into the upper stratum, and there construct "spread" footings; but it was a most unsatisfactory, and at times uncertain, procedure, and in fact, acted to retard building to great height—twenty stories was about all those old "floating" foundations could stand. Even then the buildings settled alarmingly, and it was sometimes years after completion that they came to any real stability. We have noted the case of the Federal Building in Chicago which was condemned on account of its excessive settling. It was taken as a matter of course in some of those earlier Chicago skyscrapers that they would settle, sometimes as much as a foot or more, and little was thought about it if they did not get too far out of plumb.

Yet when the early engineers in Chicago and elsewhere commenced to develop the foundations we are discussing, they had advanced little beyond the engineers of the Renaissance, and as in other great problems in which they were groping, the trial and error method was about their only guide. I have spoken of the "spread" or "floating" foundations used in Chicago and elsewhere, but there was considerable pile-driving too. The practicable limit of length of a driven pile is about eighty feet Straight trees of the necessary thickness do not grow much taller; and even if they did, pile-driving equipment to handle them is impracticable at such lengths. We have seen that in Chicago, under these pioneer skyscrapers, the hard-pan or rock is from seventy-five to a hundred feet down, and piles of that great length cost excessively. So the early assumptions when piles were used turned on the friction the pile would exert on the ground into which it was driven. As each pile was pounded down by the familiar old device of a great iron weight hoisted to the top of a tall ladder-like structure called the "ways" and allowed to fall with a terrific thud on the pile, capped and held in position at the bottom of the ways, the "penetration" was noted; that is, the amount the pile sank under the impact of each blow. Theories sprang up and a pseudo science of pile friction was devised. But alas, buildings had an unhappy way of settling where these formulæ came anywhere near the margin of safety. There was, of course, and still is a scientific basis for the resistance a pile will offer, but the *reductio ad absurdum* used by some of these early engineers to justify pile installations of short, and therefore inexpensive, lengths, depending largely on "skin friction" to support sizable structures, led to great disillusionments and a general discrediting of the method as it was then applied.

Whatever has been said of the skimping methods and theories is not to be construed as any criticism of the proper use of wood piles. Where hard-pan or rock can be reached and the piles cut off and capped below the permanent ground-water level, we have in wood piles one of the soundest and most dependable foundations. A sound pile, twelve to fourteen inches at the butt, driven to "refusal," i. e., where it refuses to penetrate farther under a given hammer impact, is as scientifically sound an element of foundation as any we know. Such a pile is good for fifteen to twenty tons of superimposed weight. Therefore, a column footing bearing, say, six hundred tons may be constructed of a group of thirty to forty such piles, capped with a bed of concrete, and may be compared to the strength of a sturdy table with a heavy stone top; but unlike the ordinary table, this one has from thirty to forty legs, and packed around those legs is a heavily compressed bed of sand, clay or even muck—whatever the substance through which the piles have been driven—to give lateral stability to the footing so built.

Shortly after the war I had to do with the erection of some large buildings in Japan. There we had occasion to use wood piles in the foundation. The soil of Tokio was ideal for the use of wood piles, being saturated at all times by the ground water from the tides; for an absolute prerequisite for the use of wood piles is that, when completed, the pile foundation must thereafter forever be submerged. So we imported great piles from Oregon and drove them sixty or seventy feet through the soft alluvial bottom of Tokio to hard-pan. These piles were in groups of the proper number under each column footing, and as is always the case, were cut off below the ground water level and then capped with blocks of concrete. Earthquake proof? Perhaps. At least they weathered the great quake in September, 1923.

In the West, the skyscraper builders have developed a unique and indeed wonderful form called the "open caisson," adapted especially to the soil conditions found in and about Chicago, and some of the cities along the Great Lakes. There the soil formation consists of a loose, wet muck and quicksand on the top to a depth of ten to twenty feet, then a stratum of dense blue clay that grows harder and denser as the depth increases. The process consists of nothing more than sinking wells, anywhere from six to ten feet or larger in diameter, according to the engineering requirements, one for each column of the skyscraper to be built. Through the first twenty feet or so it is tough going, with water and muck, and sometimes quicksand to contend with, but we do it with short plank, like straight barrel staves, about five feet long. The men dig a hole in the exactly determined location, stand up these staves,— "lagging" it is called,—have ready collapsible iron rings that are sprung into place on the inside of the cylindrical hole so formed to resist the inward pressure on the lagging. These rings correspond to barrel hoops, but are located inside instead of outside. The tendency is for the earth to crush the caisson so formed, and the iron rings prevent this. Having set one section of lagging, well braced, another five feet or so is dug and another circle of lagging with the supporting rings is set and secured. The operation is repeated, down, down, down, until a

Courtesy of Underpinning & Foundation Co., Inc.

After a tube has been driven a certain depth, a blast of air under pressure of 100 pounds per square inch is suddenly released. The explosion forces the material within the tube out of it.

Courtesy of George A. Fuller Co.

An open caisson with interlocking edges of steel sheet piling. They are made in various lengths, and when driven, present a continuous barrier against quicksand and water. Such a cofferdam is cross-braced at intervals as it is driven, and the material excavated from the inside.

Courtesy of the Foundation Co.

The enormous supply of compressed air necessary for a large caisson job is indicated by the machinery shown above. This plant was temporarily on the job for the American Telephone Company Building, New York.

deep well, eighty to a hundred feet, is constructed,—a series of straight-staved barrels one on another, but without heads, thus forming a deep well. When suitable bottom is reached, concrete is sent down in buckets and the well is filled, making a solid column of concrete from extreme bottom to the proper height where it comes up to receive the column base, generally just under the basement floor level. The thing that makes all this possible is the stratum of hard blue clay which is impervious to water; indeed, water has to be sent down to lubricate the shovels of the diggers. The water in the soil at the top two or three sections of lagging is excluded by caulking the joints, thus sealing them; a tedious operation and one that could not be made effective without the bottom-sealing effect of the blue clay which prevails through the lower and greater part of the operation.

If we are watching a Chicago skyscraper operation, we may see a curiosity that may even seem amusing. One whole area of the site, perhaps all of it, seems to be unusually well planked over, with small, framelike arrangements placed about over this area in surprisingly regular rows and covered with little tents, for rain must be kept out of these operations. Connecting these frames there is a continuous rope or belt driven from an engine or motor and engaging a pulley at each frame. And on the shaft of each is a small winch or "niggerhead" that looks like a steel spool. Wrapped a few turns around this spool, but with the spool running freely in it, is a manilla rope attended by a watchful man, who seems very much interested in peering down into a hole about two feet square under the centre of the frame on which this pulley and spool operate. At a signal from below, the man tightens his pull on the now idle rope, and immediately it commences to take hold, to wind up on the engaged niggerhead. It is a principle of friction as old as mechanics and still as useful as when it was first applied. What is happening is that each of these little frames is the heading of a deep open caisson. The man controlling the rope is looking into a well, brilliantly lighted with electric lamps at intervals of its depth. His signal from below announces that a dirt bucket has been filled, and up it comes, is swung aside and unceremoniously dumped on the platform, where the waste material is gathered by a passing wheelbarrow-man serving a number of such holes. The bucket is as unceremoniously lowered by the attendant paying out on the slack rope in his hand, the niggerhead turning on unconcernedly in its uninterrupted trundling, day in and day out, until the caissons are finished and the little rig borne off, perhaps to serve on some other waiting job, where again this set of rigs takes up another task of pulling deep cores of clay out of the foundation of the city, that stronger foundations of concrete may fill these holes and thereby make stable a mighty skyscraper.

Engineers are a proverbially dissatisfied lot and are everlastingly trying out new ways of accomplishing things around buildings. This Chicago method of open caissons was made possible, as has been said, by the underlying stratum of blue clay. The ease of the method has prompted excursions by foundation engineers into other soils, not so tractable, but a way has been found in many cases to make them surrender.

Take quicksand, for example; a most treacherous and unstable material with an evil reputation from time immemorial. Now, quicksand is just a very fine sand with an admixture of mica, pyrites and other light, sandy substances which, when saturated with water, has the property of giving way under any considerable weight placed upon a part of its surface. The weight slowly sinks into the quicksand. The fineness of the sand, together with the buoyancy of the schist, creates a condition where every particle is surrounded by water and the aggregate is, in effect, a liquid. Where the particles are larger and sharper, they touch and interlock with each other, and we have the common phenomenon of a non-liquid aggregate such as is to be found at the sea and lake beaches.

Dangerous and treacherous as quicksand is, there is a lot of solemn nonsense about its unconquerability that has challenged engineers to conquer it The Chicago caisson was the bait, for quicksand without water is as stable as any other sand. Thus, under some very important structures, foundations in the past few years have been built in and through this supposedly impossible material by the simple expedient of first sending down drain pits ahead of the pier excavations. These pits must be skilfully constructed so as to admit the water, yet exclude the fine particles of sand. Connected with these pits are pumps that have capacity to pump out all the water that runs into the pits, and these pumps must be kept running night and day until the foundations are finished. Two or three of these in a lot of the average size will draw off the allied water and leave the refractory sand tame and tractable. With the water drawn off, the open caisson can be used, of course with modifications of method to suit the material encountered.

The thing is not quite as easy as it has just been described to be, and only skilful and experienced engineers are to be trusted with it. For example, the most careful reconnaissance and soil examination must be made, with tests and borings so that the bed-rock or hard-pan will be there just where it is expected to be. It must be known whether the bed is extensive or simply a local pocket, and whether it overlies a fairly level and homogeneous hard material Generally steel sheet-piling surrounding the whole area of the building must be driven, the lower edges either sealing into the underlying bed or otherwise insuring that, with the drain pits constructed, they will surely take care of all the water, and particularly the danger that arises from the surging upward thrust of an unexpected flow of water from below, perhaps from crevices in the rock or hard-pan.

So it is seen that foundation engineering is something more than mere technical training. For the design of superstructures, such training can often be used with a modicum of experience added, but in foundations for great skyscrapers we must look for the top-notchers, the men of technical training and wide experience in the application of their ideas, and, with these qualifications, the quintessence of all values in human knowledge—abundant common sense.

Concrete piles are an adaptation of the principles of reinforced concrete to pile making. Many of the limitations of wood piles are obviated by these useful devices. Structurally, a concrete pile is a combination of cement and reinforcing rods and hoops so placed as to give maximum strength and stability to the finished member. Cast and allowed to harden, they are driven much as wood piles are driven. They are, of course, indestructible and do not depend on any particular water level for their permanence and, being vastly stronger than wood, greater weights may be imposed upon them. Like wood piles, they are the surest when they can be driven to hard-pan or rock, and also like wood piles, they are not to be used where the soil through which they are driven contains boulders which deflect and break them before they can reach sure bottom. Concrete piles over a hundred feet in length have been cast and driven

An interesting form of pile foundation is the Raymond Concrete Pile. This is generally used where rock or hard-pan is very deep and where the increasing density of the soil warrants reliance upon the skin friction of the pile. These piles may be made as long as thirty-five to forty feet. The ingenious principle involved is the making of a series of tapered casings of thin sheet iron with a spiral "hooping" within the casing, which serves both as an element of rigidity in the empty casing, and also as hooping in the concrete with which the pile is later to be filled.

The driving is done by means of a driver of sufficient height to permit of starting the pile at its extreme length. The actual driving is done by a trip-hammer, which "rides" the pile down. The operation is started by raising a heavy, collapsible steel mandrel or pile core high up into the guides on which the sections of casing are slipped from the bottom. The mandrel is tapered, and therefore, the largest section of the casing is slipped on first. This is done by the simple means of a little tackle from the top of the guides which jerks the casings up in place, the tapering sections being slipped on quickly in succession. The mandrel is thus covered with the sheet-iron casing with its hooping in it.

Now the pile driving starts. The first few blows generally seat the pile, as the upper stratum of earth is apt to be soft, and from there down, the pile is swiftly driven, the trip-hammer following the pile down in the guides of the pile-driver. As the end of the pile reaches the hard material, some twenty to thirty feet below the surface, the penetration of each blow of the hammer becomes less and less, and finally the pile comes to "refusal." Now the tapered core of the collapsible steel mandrel is withdrawn about a foot, which of course loosens the sections of the mandrel so that it is loose within the casing. It is immediately hoisted out, and we have a completely cased hole of the desired depth and into hard soil. This is then filled with concrete, and within a few minutes after the driving is started, we have a concrete pile which only awaits the setting of the cement to make it a hard, indestructible and thoroughly adequate element of the foundation.

The pile-driver on heavy rollers is easily shifted from place to place and can be accurately adjusted to position for driving each pile in its determined location.

A group of these piles, as of piles of any other form, when capped becomes a column or wall foundation. The penetration of the piles under the last few blows of the hammer, together with their known skin-friction, gives an accurate basis for determining the carrying capacity of the group. The method produces as permanent and sure a foundation as any other when properly directed by skilful engineers.

Steel tubes for foundation work have been wonderfully developed and have solved for the engineer a number of very troublesome problems. Where the soil contains not too many boulders and where hard-pan or rock may be reached with reasonable certainty within fifty feet or so, tubes are driven in clusters much as wood or concrete piles would be, but improvement in the method of driving it has made tubing possible. Reference has been made to the old fashioned gravity pile-driver that raised its great weight and let it fall with a thud on the pile That kind of driving is no longer seen around a metropolitan skyscraper foundation. Instead, we have the steam or pneumatic trip-hammer. It is a heavy engine-like device that rides on the head of the pile or tube to be driven, and can be recognized by its rapid, staccato pounding as it carries down the member it is driving. Its aggregate impact per minute is more than that of the laborious old pile-driver hammer, but the impact is distributed and we avoid the bursting force of the single, irresistible blow of the old weight.

In a tube foundation, the tubes are accurately set as piles would be; then the driving starts and the tube penetrates, say, ten feet or so. The tube is twelve to sixteen inches inside diameter, according to the requirement of the operation in hand. As soon as the driving resistance for the tube halts its downward progress, the driver is removed to one side and the derrick drops into the driven tube another tube about fifteen feet long, of a size to slide in freely. This inner tube, sometimes called the agitator, has connected with its top a small three-inch pipe which again can be passed down inside of it, and connected with the upper end of the three-inch pipe is a hose leading to a large tank of compressed air, kept at about a hundred pounds pressure by a compressor generally located on the street or in some place out of the way of the foundation work. The agitator looks like a very long bucket, the bail of which is held by the fall of the derrick. The bucket is, of course, bottomless. The derrick now commences to pump the agitator tube up and down, and as it strikes bottom in the driven tube, a man at the control air valve opens it suddenly, thus causing almost an explosion in the agitated earth within the tube. The effect is to send a perfect geyser of mud and earth flying up out of the agitator tube. Up and down the derrick jumps this tube and the effect is to clear the driven tube of everything inside of it. Again the driving and again the clearing process goes on. As the driven tube goes down, there is frequent inspection of the bottom of the hole as fast as it is cleared by the air jets. This is done by the simple device of dropping an electric light down inside the driven tube where the character of the soil may easily be observed. When the driven tubes strike rock or the predetermined kind of bearing, they are cleaned out for the last time, the bottom inspected, and if found satisfactory, they are ready for concrete. The predetermined number of these tubes having been driven for a given footing, the projecting upper ends of the tubes are cut off by an oxy-acetylene gas torch to the desired length, and the driven tubes are filled with concrete. Thus we have a number of solid columns of concrete, each surrounded by a steel casing, extending from rock to the under side of the fundation block. The groups of tubes are capped with a single block, as in the case of either wood or concrete pile groups. The rock or hard-pan is seldom level, hence the varying lengths of the driven tubes and the necessity for cutting the tops to an even level to receive the footing cap.

Sometimes these tubes encounter boulders on the way down, and if these cannot be broken up by the agitator, there is nothing left to do but dig down beside the tube and remove the boulder by derrick, if it be of moderate size; if it be large, it must be drilled and blasted. For this reason it is important to know by preliminary borings and tests whether a bouldery soil is to be encountered. If boulders appear to be too numerous, it is almost sure to be more economical to go to some different method of foundation, for digging for boulders is an expensive procedure.

This tube method is more useful in underpinning adjoining buildings, for it can be done with such little head room and in such cramped quarters. It has become the most general method for this adjoining work, and even when pneumatic caissons are used in the foundations of the new building, the adjoining ones are frequently underpinned with tubes In such cases, where it is not possible to get headroom even for a pneumatic hammer, tubes are used, jacking them down in short sections and using the old wall as a base against which the powerful jacks press the tubes downward. Special tools are used to clear the inside of the tubes. As each section is jacked down and cleared, another section is placed on top of it and the process repeated until the desired bottom is reached. The couplings used between sections are ingenious internal sleeves, the use of which avoids the necessity of outside couplings such as would be found in ordinary pipe work. One after another these tubes are placed, the work going on at several places along the wall simultaneously, until the entire old foundation is caught up on a large number of tubes and the old structure secured at the new depth. If the soil beyond the old foundations tends to cave in, horizontal sheet piling is driven in behind the tubes, which will retain almost any kind of soil but quicksand. When this is encountered, special ways and means must be devised, and it may result in the necessity of underpinning two or more rows of footings of the old building before the new foundation can proceed with safety.

The steel tube foundation, whether for underpinning or for new foundations, has been a wonderful forward stride in the everlasting foundation problem. One of its variations has been the development of the "pretest" foundation, an ingenuity that could only have been developed out of the complexity of modern skyscraper requirements of swift operation in the very limited and cramped quarters that the building site usually affords.

The pretest pile is an adaptation of the tube where it is desired to use the weight of the rising structure in course of erection, thus enabling the work of superstructure actually to precede the building of the foundations themselves. The work is carried on by starting the concrete bed of a predetermined size at the grade and in the exact position that would be indicated for a capping of any group of piles. The column base and columns are set on this in the usual way as steel erection commences. As the load accumulates, say seven or eight stories of the building, excavation is made under a part of this prepared concrete bed, and in a little temporary chamber there constructed, with about four feet of headroom, powerfully hydraulic jacks are set up. A short section of steel tube three or four feet long is set down into the ground just as an ordinary tube would be started. These are pressed down, one on top of another, the interior being cleared out as the pile is sunk, just as would be done in the case of a steel tube in the open. When this composite and now continuous tube has reached the desired bottom, it is filled with concrete and capped, and the pressure which the jack has been exerting against the under side of the footing is taken up by a permanent section of steel I beam or other heavy member securely wedged. The pressure on this pile when completed is accurately noted through a pressure gauge on the hydraulic jack, and since it is to be one of a group of, let us say, ten under this particular column, with the ultimate load on the column accurately determined, it is a simple matter to load on this single pile more than it ultimately will carry as a part of the group.

Having thus driven one pile, another and another are driven, each one carried to the desired bottom and each one tested to an overload. Thus, the accumulated carrying capacity of the group of ten has been established by pretesting it, one pile at a time, and the combined carrying capacity of all of them together is in excess of the ultimate loading of the column. It is customary to load each pretest pile to about fifty per cent over the estimated requirement and thus a combined pretest footing will be, in the aggregate, fifty per cent over the requirement—a perfectly safe and highly ingenious foundation.

Of course, in order to carry out this method, there are certain prerequisites. The soil on which the original concrete base was laid must be secure enough to carry the partial load of the structure before the pretest operation starts. However, even in such cases, this pretest work has been done by actually building the working chamber in advance of the footing block, making it on a concrete form much as a floor arch is made. This block, with the hollow chamber under it, rests its outside edge on the ground and on the water-tight sheet piling forming the chamber, much as a saucer might set on a cup. Here the short pretest sections are set up, even before two stories of steel have been set. All that is required is enough superimposed weight to exceed the loading that will be required on a single pile. In this case the jacking down is simply started earlier, and as fast as the weight above accumulates, new pretest tubes are started until the ultimate number under that footing has been completed.

The invention and use of steel sheet piling has contributed as much as any one thing to the solution of the deep and difficult foundation problem. In the earlier and simpler founda-

tions, it was common practice to drive sharpened plank behind wales to hold the earth. But if the earth were wet and
soluble, it was soon found that it ran through the crevices between the planking. Then tongue and grooved sheeting was
introduced with some success, but leaving much to be desired.
Both methods were very limited because the plank of either
type would be battered to pieces at the top before the driving proceeded ten feet or more. All sorts of caps and heads
were devised, but the uncertainty remained—one plank might
drive ten or twelve feet, the next one split at half the
depth.

Steel sheet piling could be made in any desired length; it
was tough, and the ingeniously locked edges furnished both
a guide and a bond for each succeeding piece as it was driven.
The trip-hammer principle of driving, introduced at about
the same time, was ideal for this new material, and at once
deep sheet piling immeasurably simplified deep foundation
work. One of its greatest advantages is that it is almost automatically self-sealing Trickling sand and saturated earth drip-
ping through the joints leave deposits which, in a few hours,
seal the joints, and we thus may have a continuous and impervious sheet of steel driven to almost any desired bottom before the excavation is fairly under way. Properly braced with
wales and shores, as would be any other form of sheeting, it
almost ideally fills a requirement without which many a deep
foundation would have to be abandoned as designed, and
some vastly more expensive method of coping with the wet,
unstable soils devised.

The spectacle of a great building structure going up before
the basement is excavated is not uncommon, especially in cities
of the West, where a homogeneous soil and the open caisson
foundation make it sometimes inconvenient to carry on the
excavation work while the caissons are being sunk. When this
procedure is followed, the builder disturbs the soil as little as

possible. He builds retaining walls in trenches, thus using the existing earth to brace the sheeting that holds the streets. Similarly, the adjoining underpinning is done in trenches, and all the while the caisson wells are being sunk. If the basement is to be a deep one, with ground water to contend with, the retaining walls are water-proofed as they are built and the membrane of water-proofing carried through under the wall. This is done by first laying the footing of the wall, which generally is constructed with a ridge of concrete along its centre, something after the manner of the tongue on the edge of a tongue and grooved plank. The trench for the wall is dug about a foot beyond the line of the outside face of the retaining wall, or if it is sheet piled, the wales and sheeting are so placed as to leave absolutely unobstructed a space of about a foot outside the wall. Having finished the wall footing with its ridge or "key," a wall of hollow tile is laid, the inside face of which forms the outside face of the retaining wall. Before the retaining wall is started, a membrane of hot tar and tar paper—roofing felt—generally five-ply, is laid over the footing with a projection of about a foot inside the inside face of the wall. This same membrane is carried up the inner face of the hollow tile, somewhat as a paper-hanger papers a wall. When this work has been finished to a convenient height, the concrete form at the line of the inside of the wall is set and the concrete is poured to a level about a foot below the top of the membrane, the tile wall with the membrane on it forming the outside form. Again the hollow tile is laid up scaffold high, again the tar and felt, and again the concrete, until the work is up to grade. The effect of this is to form a continuous membrane on the outside of the retaining wall and under it, with a foot or so protruding into the basement, to which will be attached the continuation of this same membrane under the basement floor when it is finally laid. Similarly, the membrane to the top of the wall is so devised that, when the side-

walk structure is set, the membrane can be joined to the
water-proofing membrane that will underlie the sidewalk. As
the caissons are finished and concreted, the membrane is carried under the grillage-bed or column base, with edges of a
foot or so to be joined with the future basement floor membrane. Under these columns where the concentrated loading
is so great, the membrane takes the form of copper pans or
sheets, carefully soldered and sealed.

We have noted the key on the foundation of the retaining
wall. This acts as a brace to keep the wall from sliding inward
when the pressure of the bank from the outside comes on the
wall, for the membrane, of course, prevents any bond between
the retaining wall and its footing. This process of constructing
the membrane in a number of places, all eventually to be
joined together, is carried on whether the excavation is completed before steel erection starts, or after. The effect in either
case is the same. As the basement floor is prepared, a section
at a time, this membrane construction is laid on a bed of concrete placed to receive it, and upon the membrane, the body
of the basement floor structure is laid. From each section pro-
trudes the necessary lap of the membrane, and when the work
is completed, the effect is a continuous sheet of everlasting
membranous water-proofing, from a point about a foot high
back of the water table of granite or marble on the building
line at the street level, out over the sidewalk, down the outside face of the retaining wall and under it; joining with a
continuous sheet underlying the basement floor and columns
and up to the height of the ground floor on the outside of the
rear walls. Thus the basement is sealed as tight as a ship's
bottom against seeping water, the membrane buried behind
the protecting concrete of the basement floor and wall structure.

With the membrane thus completed, it is common practice
to have in the lowest basement, generally the boiler room, a

sump or cistern dug six to ten feet below the floor level. This membrane passes over the upper wall of the sump and is sealed to the curb around the sump. In the soil under the concrete slab on which the membrane was laid are numerous lines of agricultural drain tile that radiate in all directions from the sump and slope slightly toward it. The ground water strikes these drains and flows to the sump, and pumps that start automatically as soon as a certain predetermined water level in the sump is reached, lift this water out to the sewer. In this way, the upward pressure on the basement floor is relieved. If the pumps fail to work for any reason, the water simply overflows the sump and the basement floors, but the water-proofing membrane is not subjected to pressure from beneath.

When sumps are not used and the water pressure is considerable, it then becomes necessary to reinforce the floor structure overlying the membrane. Sometimes this calls for a concrete overstructure two to three feet thick, and then reinforced rods,—a reverse reinforced arch construction. A sealed basement of this kind forty feet below ground water level will receive an upward thrust of nearly eighteen pounds to the square inch. A little calculating of the aggregate upward pressure on a basement of, say, ten thousand square feet, will indicate the tremendous force that must be neutralized if the sump is omitted.

There is a method of excluding water from basements by plastering the inside of the basement walls and floor with an impervious coating of water-proofed cement, which takes hold of the masonry with such tenacity as to become integral with it. Such a system must likewise be continuous, but instead of passing under the columns, it carries up on the outside of the surrounding masonry with which the columns are fire-proofed, generally they are encased in concrete to receive this water-proofing. In this method, columns are in effect standing in sleeves of water-proofing themselves—theoretically, at least—standing in the saturated concrete that encases them.

Returning to the excavation problem that is sometimes met by taking out the general basement earth after the structural steel has been set and the retaining walls built, it has the great advantage of requiring a minimum of shoring and bracing. As we have seen, the streets are temporarily braced against the opposite bank of a trench, and the problem of interior shoring, where there are different basement levels, is much simplified. When this method is followed, columns are lowered into place virtually in pits, and field engineering to set the column bases is most difficult, but skilful builders find no trouble in doing this. The columns having been set, the floor beams are placed and sometimes the floor arches filled in. In this way, the permanent bracing of the retaining walls is effected without the intermediate steps of shoring and bracing. After the steel has been set, the excavation proceeds, sometimes by means of specially designed, low-headroom, power shovels, sometimes by hand loading. Whatever the method, provision for this post-construction excavation has been carefully thought out by the builder, and whether he makes provision for bringing trucks directly into the excavation or hoists the material out by derrick or hoistway, the result is that the superstructure proceeds while the excavation is being finished. The character of the soil, the expediency and the imminence of steel delivery all have a bearing. It is one of the complexities of the builder's problem, and to say that it is sometimes done is all that can be said as to why it is done.

I remember one deep foundation we had to put in, where we were obliged to go down through the floor of a large engine-room. The engine had to be kept running to generate electricity for a great existing skyscraper to which we were adding. The engine and dynamo were on a heavy bed of concrete about ten feet thick, and a part of this had to be cut away to let the new steel column pass beside it to the new deeper footing. Water unlimited flowed in the gravelly soil below the engine bed. And there we worked night and day, sinking the steel sheet piling of a cofferdam in that engine-room floor, within a few feet of the fast flying crosshead of that great machine as it whirled back and forth, the dynamo delivering uninterruptedly its vast power output without so much as dust from our work interfering. When at last we reached hard-pan thirty feet below and commenced placing the concrete, we felt the elation of a battalion commander who had carried a redout and signalled the regiment to follow. Then came the job of setting the base for the column—as nice a piece of field engineering as was ever done; and when the great steel column, weighing about ten tons, was gently lowered into place and bolted to the base and found to be exactly in position with respect to all the other columns, we knew we had won a great victory of peace—the accomplishment which makes us know that building is an inspiring calling. The column stands to-day in that basement engine-room, with others, looking like the pillars of Hercules; but to an intrepid band of workers, it is the monument of a great victory.

And so we have discussed at great length the matter of foundations, for foundations are all important in skyscraper construction, and were it not for the labors and ingenuity of those pioneer engineers and builders in devising safe and sure foundations, the height limit would long ago have been set by the limitations of the old-fashioned footings. Again we must say that this development is one of our own creating, devised to meet the needs of our most characteristic American accomplishment—our beloved skyscraper. All of this is the work of the builder; digging down into the depths to build the mighty foundations that he may turn back and, with his skyscraper, conquer the towering heights. "Laying a good foundation" is the metaphor universally used. It connotes the proper starting of any good creative effort, and from the builder the metaphor is learned.

Courtesy of Thos. Crimmins Contracting Co.

A caisson job where there is little space for concrete mixing. Concrete being delivered by special revolving-drum truck which keeps the concrete mixing as it moves through the streets.

CHAPTER XIV

STRUCTURAL STEEL IN THE MAKING

WHILE all this activity of foundation building at the site is going on, a hundred things are happening in the office of the skyscraper builder. The structural steel has been ordered, and shop drawings by the ream are being approved and blueprints of them forwarded to the bridge shop, where men translate them into completed structural members. The bridge shop has furnished the rolling mill with lists of plain shapes and sizes cut to accurate lengths, and it is the business of the bridge shop to take these plain pieces, punch them, rivet on the lugs, build up the columns and girders; all exactly in accordance with the shop drawings. As the pieces are finished, they are marked and numbered in accordance with the setting plans, prepared by the structural engineer as the common guide for all concerned in the design and erection of the steel. For the structural steel is the pivot around which the whole superstructure of the skyscraper under construction turns. Excavation and foundation are timed to the delivery date of the steel, and all plans for enclosing the building must depend on the steel erection.

It may be of interest, before watching this, the most spectacular of the operations of building, to make a brief excursion into the steel mills and see how steel is made, and into the bridge shops and see how it is fabricated, that we may better appreciate the wonderful development that this national demand for great structures has brought about. For it is true that our national genius for construction, which has no parallel in all history that can even remotely approach it, has largely been made possible by the amazing perfection and unlimited capacity for production with which the steel industry has led the way. Construction and steel production are inseparably linked;

neither would have been possible without the other, for the demands of the one furnished the incentive for the colossal scale upon which the other has been developed.

Without attempting to start too far back, we will begin with the blast furnaces at the Bethlehem Steel Company, where the base ore receives its purifying with a proper mixture of limestone, manganese and other ingredients that the metallurgists know, for these ingredients are the scavengers that free the molten metal from impurities, sulphur, excessive carbon and the like. The tremendous and continuous heat of the blast furnace is furnished by coke with an unbelievable amount of air blown through it in order to furnish the excess of oxygen necessary to produce the heat. The ore, limestone, manganese, coke, all are piled in the furnaces and ignited, and after several hours of furious heat, the molten mass is drawn off and cast into pigs —pig-iron. The pig-iron goes to the open hearth furnaces, where it is again melted and receives the final purifying and the ingredients that give it the desired metallurgical properties of steel.

In the great modern steel plants, each of these open hearth furnaces produces about eighty tons of steel at a single heat, and the temperature maintained in them is nearly three thousand degrees Fahrenheit. These furnaces are in a row down the length of the great shed by which they are enclosed, and along this interior, in front of the furnaces, runs a heavy crane-way with cranes of prodigious lifting power. The heat being ready for "drawing," a spout is opened in the side of the furnace, a great ladle having first been placed in position by the crane to receive the flow of metal as it pours out of the furnace. There another receptacle is placed alongside and so arranged that a spout from the top of the molten mass will carry off the lighter material, the molten slag from the top of the great stream of metal. The slag itself has valuable uses, but we will not follow them here beyond saying that it furnishes one of the necessary

Tapping an Open Hearth Furnace. Molten metal is starting to pour into the huge ladle.

The filled ladle is about to be raised by the huge crane. The lift is about 100 tons. The thimble-shaped vessel on the side receives the molten slag, which, being lighter than the metal, can be drawn off by a separate spout from the flux as it pours out of the furnace.

ingredients of Portland cement—that equally indispensable need of modern construction.

The ladle having been filled, the crane raises it and whisks it away as though it were out of the power of gravitation, although it carries a burden of eighty tons of molten metal. Arranged down the shed on specially constructed cars are great moulds the shape of a billet, for the metal is now on its way to be cast into billets of from six to fifteen tons or more each. The crane stops over each mould in turn and a gate in the bottom of the ladle is opened, filling the mould with molten metal, much as a housewife fills a muffin tin. The train-load of moulds having been filled is moved away to an adjoining position for the next operation, known as stripping. These moulds are slightly tapered to facilitate this step for, as the metal solidifies into red hot billets, the moulds are passed under a crane that carries the stripper—a mighty pair of jaws that engage the lugs on the moulds, and while they lift the mould from the billet, they are supplemented by a great plunger that presses down and shoves the billet through the mould if it tends to stick, and the red-hot billets are left standing on the cars, their moulds stripped off. The train-load of red-hot billets is run into another shed and again a great crane hovers over them, this time to set them gently into the "soaking" pits. The original steel-makers must have had a sense of humor to have devised the name, for soaked they are, but with an inferno heat from a roaring gas flame that raises and evens the temperature to a point about two hundred degrees below liquefaction. And now the billets are ready for the next operation.

From here the crane carries the white-hot billets, now thoroughly "soaked," onto the power-driven rolling bed, and immediately they start on their way toward the rolls. It is one of the great spectacles of industry to see this mighty drama. The billet—perhaps ten tons in weight, two feet by three feet across and about seven feet long- —comes tearing up to the rolls car-

Pouring ingots. The molten steel is seen flowing from the ladle into the ingot or billet mould.

Stripping ingots. The ingot moulds are lifted from the ingots by the stripping machine. The great plunger in the stripping machine presses the ingot through the moulds, which are slightly tapered to facilitate freeing the ingots.

ried by the whirling, power-driven roller bed on which it
rides. The rolls themselves are whirling and of prodigious
power, great spool-like affairs perhaps three feet in diameter,
one set above and one below, with great bearings thicker than a
man's thigh and held in a ponderous frame which admits of an
adjustment of the rolls closer and closer together as the process
proceeds. These rolls grab the billet and squeeze it through.
Similar rolls in a vertical position in the frame are so placed as
to limit the side spread of the billet as it is squeezed forward.
Stationed above the rolls in a bridge-like housing and intent-
ly watching the action, is the roller who guides the operation.
He and two assistants, one stationed with him, and one stationed on a platform beside the machine, control all of its
movements. As the great billet, still white hot, comes squeezing through the rolls, the bed rolls are immediately reversed,
and back it goes to be squeezed again by these same unrelent-
ing rolls, this time set a trifle closer together by the guiding
hand on the bridge above. Again the reverse of the bed rolls,
and again it goes racing through, this time appreciably elongated. The rolls are about twice as wide as the billet when it
first came from the soaking pits, and it is noticed that in its
first few passages back and forth, the rolling is done on one
side of the rolls, the other side being of considerably larger
diameter and grooved. Now, as the billet passes through for
the fourth or fifth time, a surprising movement in the bed of
the machine makes a great clamp jerk the billet aside, and this
time it travels back through the part of the rolls with grooves.
Here the flanges commence to form and the billet begins to
assume a decided elongation; but the machine is relentless and
as the great bed rollers hurl the billet back and forth, again
and again, forcing the now elongated and nearly formed mass
of steel, it seems fairly to writhe in agony at the gruelling
process. All unexpectedly to the spectator, the elongated piece,
now perhaps sixteen inches in cross section each way, and

The white-hot ingot has started on its journey through the power-driven rolls. Their action and spacing as the ingot is hurled back and forth are controlled by the chief roller located in the "pulpit," the bridge over the rolls.

Courtesy of Bethlehem Steel Co.

Here the ingot has about capitulated to the elongating process. The rolled section, in its final form, shoots out into the cutting sheds.

twenty feet long, with the depressions that form the web of
the section we are watching, rushes forward toward the next
set of rolls, for the first set is done with it and another hot
billet is waiting.

As the now elongated billet passes ahead, it runs under a
mighty shear, also man-controlled, and as its front end goes
in a few inches, the bed rollers stop and the shear closes, cutting off the "bloom" end, for in this furious squeezing process,
the ends have become dog-eared and deformed. Moreover, the
last of the impurities of the steel are apt to be found where
they would have floated to the top of the mould when the
liquid steel was poured in. The shear cuts the white-hot metal
as neatly as a confectioner cuts a column of hot molasses candy.
Again the great mass moves forward; again the shear at the
other end; and on the piece whirls to the next set of rolls,
where the process of forming it continues by rushing the now
elongated column or beam back and forth through the rolls
until it has been given the desired section. The whole action
takes place in a very few minutes. Perhaps twenty times back
and forth through the first rolls, then the shearing, and about
ten times through the final rolls, and on it goes, now about
seventy-five or eighty feet long, a great cherry-red column, on
the power rollers, to be cut to length.

Streams of water pour on the rolls to relieve them of the
tremendous heat of the white-hot billet as it rushes back and
forth in its process of formation and elongation. If it is a beam
section that is being rolled, the great, single piece as it comes
from the last pass of the rolls may be over a hundred feet long,
as yet unidentified for any particular structure. But whatever
the section, its makers are not yet through with it. As it passes
out of the last rolling, it is hurried on to the straightening
bed, and the moment it arrives there, the jaws of the straightener take charge. Here again an operative on a bridge over the
bed presides and, with true eye, manipulates the jaws so that

A punching operation on steel structural shapes.

A close-up view of a riveting operation in fabricating
steel columns.

Interior view of bridge and fabricating shop at Bethlehem's Steelton Plant, Steelton, Pa., showing large

the now gray-hot member is rendered as straight as a taut
string. It came into the vise in the horizontal position in which
it was originally rolled. Another lurch, the vise opens a little
way again; a quick throw of the straightener, and the great
snake is stood on edge; in a flash the vise closes again and the
great, continuous member, only a few minutes ago a white-
hot billet, is now a full fledged I beam or column section, the
last of its cherry-red disappeared, the naked gray color of the
steel preceding its final surrender to the inexorable forces
that have just fashioned it.

But on it must travel to make way for others that are following. The vise opens and kicks the gray steel length out.
Again the rollers have it and it is passing out to the cutting
room, still too hot to touch, but in the full power of its new-
found rigidity, and now very straight and true. In the cutting
room, still on its bed of rollers, it is passed into the hands of
the cutters, who hold the rolling lists from the bridge shops.
Here the seal of its identity with some particular building is
put upon it, for it is shoved onto a table where great power-
driven circular saws cut it to the exact lengths that the bridge
shop orders call for, the exact lengths that some structural engineer away back in a distant city had worked out as requisite
to fit in a certain place to be a part of a building that an architect with whom the engineer was collaborating had designed.

The rolling lists contain a large number of pieces of the
same length and size; those floor beams, for example, that
occur panel after panel and floor after floor, all exactly the
same length and requiring the same shop work. Similarly, the
girders, heavier members into which the beams frame, may be
duplicated a number of times, particularly if the building covers a considerable area and is on a rectangular lot. As the
building rises in height, the columns diminish in sections for
the obvious reason that they do not have to bear the accumulated load that the columns in the lower stories bear. Hence,

as the columns diminish, the successive tiers of girders become longer by fractions of an inch; they have to reach a little farther to meet the diminishing column sizes. The precise calculations of the structural engineer have taken all this into account in preparing the shop drawings and the accurately checked rolling lists account for every piece.

And now we have a great pile of floor beams cut to length and ready for the bridge shop; the mill is through with them.

The steel mills roll a great variety of shapes and sizes to meet the complex requirements of modern structural design; at Bethlehem mills I beams from four to thirty-three inches in depth, and in many of these standard depths there will be several different weights per foot. Thus we have 12″–25's, 12″–32's, 12″–40's, etc., meaning that a given depth of these beams may weigh 25, 32, or 40 pounds per linear foot. Besides the I beams, there are channels—beams with flanges on only one side of the web—angles of all combinations of leg lengths and weight per foot; T irons, similarly variable; Z bars, as their name implies, having one flange in each direction from the web—a seemingly endless variety of shapes, sizes and weight per foot.

An anomaly here arising out of the early independent practices of structural and railroad engineers leaves us still with railroad rails standardized on the weight per yard, whereas all structural steel standardizes on weight per foot. Thus we speak of a hundred pound railroad rail, meaning a hundred pounds to the linear yard, a very heavy rail, but it compares in weight with a structural member that would weigh only thirty-three and a third pounds per linear foot. Both are produced by the same method of rolling—generally in the same mills, and nowadays almost universally of open hearth steel.

Beams and shapes of various sizes are rolled on the same beds, but every shape demands a different set of rolls, and it is necessary for the mill to change these rolls whenever the type of shape to be rolled in a certain bed is changed. The changing involves a considerable amount of work during which that bed is shut down and unproductive. The most used sizes—8″, 9″, 10″ and 12″ beams, for example—are in constant demand, and therefore mill orders for these sizes may be filled at almost any time. But for unusual sizes—for example, 24″ and 30″ beams—the mill will await rolling until it has accumulated enough orders in these sizes to justify setting the rolls, and then produce therefrom a considerable tonnage. Hence, we have one of the first anxieties of the builder to see that the rolling lists are delivered as early as possible; and if any unusual shapes occur, so to schedule as not to miss these special rollings, for it is sometimes several weeks before these specials will be rolled again. A day's delay in delivery of a rolling list may sometimes set a steel delivery date back six weeks or more.

The Bethlehem Steel Company, several years ago, developed a column rolled in the same manner as beams are rolled. Theretofore columns were "built up"; that is, the desired column section was produced by combining plates and angles—sometimes channels, I beams and Z bars figured in the ensemble—and these securely riveted together gave the required strength and rigidity. The Bethlehem column does away with the necessity for all of this, and we are now able to obtain rolled column sections of almost any required strength. Where columns of tremendous carrying capacity are required, twelve to fifteen hundred tons and more, it is sometimes necessary to rivet plates even on Bethlehem sections. This is unusual and the Bethlehem column may easily be recognized from the street by its long, smooth shaft, free of rivets, excepting where connections for girders or splice plates occur. Splice plates, as the name implies, are those plates generally riveted to the top end of the column to connect it to the column above. If the succeeding column is of smaller section than the one on which it stands, the splice plates carry with them "fillers," i. e. steel plates of a thickness to make up for the diminished column section above

We follow this plain material into the bridge shop—sometimes hundreds of miles away, sometimes adjoining the mill, —and see the final process before the steel is delivered to the building. Here is a place of noisy riveters and clashing machinery. The plain material is unloaded from the cars by cranes and sorted to benches or beds, where it receives its punching. Again the shop drawings control—every rivet hole exactly located—every shelf angle and connection noted. Columns are stiffened by plates and angles, and girder connections constructed. If built-up columns or girders are to be fabricated, long, wooden templates with the holes accurately located are prepared, so that the punching in plain members that are to be brought together will exactly "register." Beams that frame into other beams of the same depth have to be "coped,"—that is, the flange cut away at the ends to allow them to slip into end contact with the webs of the beams that receive them. Generally speaking, a girder is any beam into which another frames for support. Girders are naturally heavier as they transfer the accumulated load of several beams to the columns or to other still heavier girders. The coping of beams generally occurs only on the top flanges, as they nearly always frame into girders of greater depth; hence only the top flanges are in the same plane. Sometimes engineers endeavor to design the framing so that the top and bottom flanges of the beams fall within the web of the girder, thus avoiding coping altogether, which effects a saving, for each operation—punching, shearing, coping, etc.—bears a charge, an "extra" or "tariff" above the cost of the plain material. It is in the economies of fabrication, the efficiency of low cost production, that one bridge shop is able to underbid another. The price of the plain material from the mill on any given job is apt to be about the same to any of several of the leading fabricating shops. It is the bridge shop that sells the fabricated material to the builder.

There is no fixed price per ton for fabricated steel in a given market, although the public has the erroneous idea that there is. A heavy job with a preponderance of heavy straight pieces and much duplication will command a much lower price per ton than a light, irregular job, particularly when special framing occurs, such as complicated sloping roofs, etc. The base price in the plain material on both jobs may be the same at the mill, but the large amount of fabrication on small and light members in the latter runs the cost per ton away up. The point must be obvious to the reader and is only brought out to show that, without knowing the character of the structural design and the relation of the extent of the fabrication to tonnage of any given job, nothing but the most general statement as to steel cost can be made about that job.

The bridge shop fabricates with greatest care, and in the seeming disorder of thousands of unassembled pieces lying about, from small bits of angle that one might put in one's pocket, to great, ponderous lengths of Bethlehem shapes, the work goes through in an orderly way and in accordance with the contract requirements of the builder. Every piece is marked for its place in the building in accordance with the setting plan. If the job covers a large area with several derricks needed in the setting, the identifying marks include the number of the derrick to which it is to be delivered. The observer may notice that if the material comes to the job with a "shop coat" of paint, that paint will leave a spot of bare metal on each member around the all-important identifying mark; this is because that mark is put on at the fabricating bench, necessarily before the shop coat of paint is applied.

Builders and engineers are frequently asked how it is possible for engineers to make these complicated calculations that govern the design of a steel skeleton. The answer is that civil engineering is an exact and learned science, and no description such as is here given could hope to compass the subject. Mathematics, through calculus and the use of logarithms, must be known. Both plain and solid geometry and trigonometry are used. The science of strength and resistance of materials is brought into use and structural designing itself deals in such terrifying phrases as bending moments, moduli, radii of gyration, sheer stress, elastic limits, parallelograms of forces and strain diagrams.

But we may glimpse a few of the essentials of the simplest problem in an endeavor to appreciate what the structural engineer might do. Take a square table with four legs and load it down evenly with a ton of bricks. Obviously, each leg will be supporting a quarter of a ton, or five hundred pounds. Put another table, similarly loaded, squarely on top of it. The legs of the upper table will be carrying five hundred pounds each, but those of the lower one will now be carrying one thousand pounds each. Repeat this ten times and we find the legs of the lowest table carrying five thousand pounds each, and each succeeding table as we go upward, five hundred pounds less than the one below it. If one were designing such a set of tables, he would make the legs of the lowest set very strong to receive the accumulated load. Now, the table tops would each be the same, calculated to support only the ton of weight each was intended to carry. If we were designing the composite of tables with care, we would simply make four very long legs and set the tops in at the proper intervals, and these legs would each be tapered toward its top. By making the legs continuous, we get a rigidity which the separate tables stacked one on another would lack. This rigidity contains the principle of wind-bracing.

Engineers know what each floor will weigh, taking into account the type of floor arch, the thickness of the floor-fill above the arch, the probable amount of partitions and, most

important, the "live load;" that is, the weight of the building
contents—people, material or merchandise. In many cities,
the required allowance for live load is forty pounds per square
foot for office-buildings, habitations, etc. From this figure
there are increased requirements for light storage buildings,
lofts, etc.—one hundred and twenty pounds per square foot is
general for this class—and from there up, even greater live
loading, until we occasionally find heavy warehouses for
storage of paper, sugar and very heavy, dense commodities,
running to six hundred pounds and more per square foot.
Such buildings, if they are tall—say ten stories or more—have
enormously heavy columns in the lower floor, and, of course,
tremendous foundations.

Wind pressure is a factor which must be reckoned with,
and the law generally makes complicated and technical stipu-
lations as to how wind-bracing is to be calculated. Reduced to
their simplest terms, it means that the law requires an allowance of about thirty pounds per square foot of wall surface
above the ground. Where buildings are tall and narrow, the
wind-bracing may be noted by the heavy gusset plates at the
intersections of columns and girders. These often take the
form of diagonal braces made of small beams or channels and
riveted to the column three or four feet above the floor and to
the girder as many feet from the column. Such bracing at the
floor can only occur in the walls or in elevator shafts as, obviously, it would be impracticable to have them on the floors in
the usable space of the building. Wind-bracing is too complex
to attempt a further description of it here. A wind of one hundred miles per hour exerts a pressure of about thirty pounds
per square foot, so it would seem that where such legal requirements are in effect, they are ample. There are no cases
recorded of steel buildings being blown over, or even blown
out of plumb, in cities where there is any pretense of building
laws. In the great Florida cyclone, which at times blew more

than one hundred and twenty miles per hour, a few buildings
were deformed, and indeed so twisted as to require stripping
away some of the masonry, so that the steel could be straightened. These were the flimsiest and least prudently designed.
All of the better structures stood like the Rock of Gibraltar,
although many of their windows were broken by the intensity
of the gale.

Builders are sometimes asked whether there is any likeli-
hood of a building toppling over even though sufficiently
wind-braced, and with this is generally coupled the question
as to whether the buildings are not anchored down. I know
of no case of toppling, even among the flimsiest of the tall
Florida structures. As to the anchoring, a moment's consideration will show such a course to be unnecessary, and indeed,
it would be futile unless anchorages so enormous as to be fantastic were provided. Try to lift the back wheels of an ordinary automobile. The weight is about a ton, and its solid stability gives no thought of any necessity to anchor it down. On
a single column of the average office-building, say twenty stories high, which comes to its foundation in about one-tenth
of the area occupied by an automobile, there is a dead weight
of perhaps five hundred tons. If the building contains fifty
such columns, multiply this stability by fifty, or a total building weight of twenty-five thousand tons. We commence to see
that old Mother Nature can safely be depended upon to take
care of all the anchoring necessary by means of gravity. The
swaying of tall, narrow buildings in a heavy wind is, however,
a reality. It is reported that both the Singer Tower and the
Woolworth Building sway as much as six inches at the top
during a heavy gale. This would be natural and, incidentally,
would imply not a particle of danger. The steel is elastic and
so is the masonry that encases it within the limits that such
swaying demands.

CHAPTER XV

STEEL ERECTION AND DERRICKS

AND with these excursions into the mill and shop and engineer's office, we are prepared to stand and watch the erection of that steel, for it is hoped that our interest is heightened by the knowledge we have gained as to its origin. Down among the foundations, derricks have been erected and it is to be noted that the forehanded builder has provided anchorages in a number of the heavy foundations to which the guys of the derricks attach. The derricks are so arranged that the reach of their booms overlaps, that they can reach into the street, and can set the most remote columns—the far corners of the building. Steel arrives, the first delivery being the basement columns, for here we are assuming that the grillage beds on the foundation piers have all been truly set and precisely in place; the work of the field engineers. Columns commence to stand up like magically produced trees, those farthest from the derrick first, then here and there a panel of steel. Almost as you watch, you notice the connecting up of the panels, and in a day or two a whole tier of beams, and as rapidly as the panels are set they are planked over.

The passer-by who stops to watch the setting of the first columns will surely feel the spirit of elation that pervades the job on this long-awaited day. A corner has been turned and the interesting steel erector takes the centre of the stage. It is a gala day, after the long siege of anxiety of foundation building. The spectator watches a few pieces of steel set, is thrilled by it, and hurries on. A few days later he passes by and the derricks are now standing on the second floor. How did they get there? The process is simple, if one but stood to watch it.

THESE INTREPID STEEL WORKERS GIVE LITTLE HEED TO THE HAZARD OF THEIR OCCUPATION.

High aloft on the steel structure of the New York Life Insurance Company Building, New York.

The riveter's bucket the only barrier between the red-hot rivet tossed from the forge and the crowd in the street, 500 feet below.

It has been noted that the derrick works by setting the farthest pieces first, building nearer and nearer itself until it is finally hemmed in, boom almost tight against the mast, and raising the last piece of steel from where it has been placed almost at the foot-block of the derrick. Immediately after this last piece is set, all hands in the erection crew turn to on the business of raising. First the boom is unseated, and the topping lift detached from the boom point and attached just above the middle of the boom. Temporary guys that are stored at hand are attached to the boom point in holes especially prepared for them, or to the topping lift ring. With this rigging in readiness, the engineer is signalled to hoist away on the topping lift, and the boom rises vertically, the loose temporary guys dangling, as yet unused. As the base of the boom reaches a point a foot or two above the floor on which the derrick is to rest, generally two floors above the starting point of this operation, the riggers slide under it heavy timbers that were placed near by as one of the last operations of the derrick before raising began. These form a temporary foot-block, and immediately the riggers carry the temporary guys out to predetermined places on the new floor and secure the boom in its vertical position. Now the topping lift is unslung from the boom, and the fall of the derrick, which has hung loose from the boom point, is attached to the mast just above the middle, preparatory to hoisting, just as the boom was hoisted Securely lashed to the bottom of the mast are the foot-block timbers, on which the derrick stands when working. The fall having taken firm hold on the mast to support it, its guys are cast off at their outer ends and all is in readiness to hoist the mast. Now the engineer is signalled to go ahead on the fall, and the mast, with its foot-block and dangling guys, rises slowly through the floor structure and above it, as the boom did. The riggers have timbers laid out to thrust into place as soon as the bottom of the foot-block is above the floor, and the mast is

Courtesy of Thompson-Starrett Company.

Setting one of the six 155-ton trusses in the Paramount Building, Times Square, New York. The lift was so devised that each of the derricks bore half the load, or about 78 tons.

Photograph by International Newsreel Corp.

A 10-ton, steel-latticed derrick after falling twenty-two stories. Happily, not a soul was injured by this accident.

quickly adjusted to its positions by the permanent guys, generally secured to column tops at the outside line of the building. When this has been done, the mast is plumbed as nearly as can be; a rigger goes aloft on a bo's'un chair and attaches the topping lift to its ring in the boom point; a short hoist of the boom until it hangs in place over the boom seat; the boom pin is shoved into place and secured, and the derrick is again ready for action.

These derrick gangs develop great skill at this operation. It has been done repeatedly in two hours, and records of an hour and a half are claimed.

Where does it all come from? Whence these planks, these rivets and forges, these hoists for material, all arrived as if by magic? It is all a part of the builder's plan, and we who watch little realize that here is the triumphant consummation of months of tireless effort of a great building organization. Yes, the things one sees and a thousand things unseen come not by magic, but as the result of vigilant and organized forethought A turning point in the job has been reached.

And now we are back to "grade" again; the anxieties of foundation and underpinning have passed, and as the structural iron-workers take the steel up to the towering heights, a very great transition takes place in that deep basement. Only a couple of weeks ago it was bathed in sunlight, swarming with foundation men and iron workers. Now it is a dismally dark cavern, the daylight shut out by the superstructure; but already the electricians have strung electric lights, and it becomes a place of sand-bins and mixing-beds, storehouses and toolhouses. A wholly different crew has taken possession and we see the new faces of plumbers, steamfitters and sheet-metal-workers. Already we are commencing to install the ventilating ducts.

But the interest is still aloft with the steel erectors, for they hold the centre of the stage as long as there are new heights to conquer. Looking up ten stories does not seem so high, but looking down—Ye gods! If there ever was an experience to bring to the human body its sense of helplessness and despair, its agonies and terrors, it is the sensation felt by one who has not had training when he suddenly finds himself out on a narrow beam or plank, high above the ground and unprotected by a hand-hold of any kind; simply depending on his sense of balance and equilibrium. I have seen men who had not had experience with height and who suddenly found themselves out on these precarious footings, get down and hug themselves around the beam, their eyes tightly shut, and gasping as though they were drowning. The fresh paint on the beam made no difference. They were oblivious to ruined and disheveled clothing; the one primitive instinct of self-preservation obliterated all others. Such unfortunates experience every sensation of imminent death, yet the devil-may-care iron-setters look on unconcerned. To them the victim's desperate plight is a source of almost hilarious amusement. Some men can never accustom themselves to great heights, and these should not follow the construction end of the building business. Most men, however, can acquire the sense of balance when aloft, but it is a thing that must be done not too suddenly. Courage must be tempered with caution. These iron-setters become accustomed as a necessary part of their daily occupation, and they are oblivious to any sensation of height. They acquire an almost reckless disregard for the ever-present danger that lurks in a misstep. The instinctive sense of balance, once attained, must be used and practised or it will be lost. Men who leave building work and return to it after an interval of a couple of years are apt to find themselves very ill at ease at a great height until they have reaccustomed themselves to it.

Steel comes to the job in truck-loads of from six to ten tons. The floor beams and small girders are bundled together and hoisted to the derrick floor, there to be sorted and then set.

Columns and girders may be heavier, and even special trucks have to be used to bring them to the site. A twelve to fifteen ton column is a very heavy one and is generally to be found in the first or second tier. Trusses may be of very great weights, sometimes over a hundred tons, where great auditoriums are to be spanned. To hoist these great weights, special derricks must be provided and much preparatory work done to reinforce the structure whereon the derricks stand. Ordinarily, however, a derrick of from ten to fifteen tons capacity will suffice. The bundles of beams one sees soaring aloft generally weigh about five to seven tons. Columns are made in two-story lengths because that is the practical limit of height for setting. Moreover, since the floor beams and columns for the two succeeding floors must be stored while being sorted on the floor on which the derrick stands, the danger of overloading that floor must be considered. Two stories of steel piled on the floor is about the allowable limit in the ordinary metropolitan structure we see.

Derricks are of great concern to the builder, and special care is taken in estimating their duties and use. The cables must be new and without a flaw, tested and of the finest make. Turnbuckles and chains are tested and frequently inspected. In the basement and for foundation work one frequently sees Stiff-leg Derricks, especially around caisson work or on the bank where deep excavations are going on. These, as their name implies, consist of two vertical, triangular frames set at right angles to each other, the mast forming a common vertical leg for each triangle. The boom is generally much longer than the mast. This derrick has the advantage that it does not have to be guyed in all directions but can be set up almost anywhere, the back end of the stiff legs weighted down with stone piles or other heavy objects. The boom swings freely in an arc of two hundred and seventy degrees. Its disadvantage is that it cannot reach any point back of itself—in the remaining ninety degree arc—excepting by the laborious and sometimes dangerous expedient of unshipping the upper leg, turning the boom back between the legs, reseating the removed leg and then guying the mast away from the new position of the boom. In its new position, the boom works only in an arc of ninety degrees.

If the reader has a flare for engineering, he might care to rough out a strain diagram with the boom working back between the stiff legs with the theoretical lift at any point outside of a line drawn between the centres of the two stone piles weighting down the heels of the stiff leg frames. Such a diagram will reveal that the tendency of the derrick in this position is to lift up at the mast—hence the necessary temporary back-guying to overcome this danger.

The Guy Derrick is the one most commonly seen in steel erection. It is now almost universally of steel "lattice" construction. Here the mast is longer than the boom, the guys, generally six in number, are attached to the "spider" at the top of the derrick, which also forms the bearing in which the gudgeon pin turns. Booms are generally from seventy-five to ninety feet long and the mast about ten feet longer than the boom. The boom hinges on the mast at its base in either type of derrick. The advantage of this type is that it can operate in a complete circle—three hundred and sixty degrees. It can also be more easily raised from floor to floor. Its disadvantage is that, except as it operates in the very limited arc between any two guys, it must "boom up,"—that is, the boom must be pulled up close against the mast before it can be turned around to any position outside the two guys between which it may have picked up its load. The process is a little tedious, but there is no practicable way of avoiding this everlasting booming in and out, and the advantages of a full arc swing are very great.

The Gin Pole is a light, inverted T shaped affair, with the

inverted horizontal bar of the tee supported by diagonal struts. This little derrick is moved about on the floor with pinch-bars to a position to set some light piece, quickly guyed and in use while we watch it. Sometimes there is a small engine on the floor to operate its manila rope fall. Sometimes its hoisting is accomplished by three or four husky pairs of hands.

The Breast Derrick, most commonly seen where stone setting is being carried on, like the Gin Pole, has a lateral motion and is pinched along sidewise to its desired position. It looks like a truncated letter A, or perhaps like an elongated Tori gate from Japan. Generally used with a manilla rope fall, it carries a winch bolted to its sturdy legs at a convenient height to enable two men to wind upon its drum the travelling part of the fall.

The Ginnywink is a light A frame derrick resembling a Stiff-leg, but smaller, and like the Gin Pole, is useful for light work where a derrick is needed in a limited area. This little derrick is also pinched about on the floor. It has no mast but its legs are like a letter A with the topping lift in the apex, the boom swinging on a pin in the horizontal sill and a folding stiff leg running out rearward, which is lashed down, as is the sill that supports the bottoms of the legs of the A. This derrick swings only in an arc of one hundred and eighty degrees and must be moved frequently.

There are many special forms of derrick, but the ones above given will indicate the standard forms most frequently seen.

These boom derricks, either stiff-leg or guy, are operated by two separate cables, the topping lift which raises the boom or allows it to lower, and the fall. The hoisting engine is equipped with two drums which may be operated independently of each other, and the hoisting engineer operates these drums in response to signals from the derrick man. As the building rises, the derrick gets entirely out of sight of the engineer, for his engine is generally left in the basement or on the ground floor.

Two enormous Stiff-leg derricks used by McClintic-Marshall Co. in the construction of the Hudson River Bridge towers. The masts are 55 feet high and the booms 85 feet long. Each derrick has a lifting capacity of 84 tons at 60 feet radius, or 65 tons at 80 feet radius. The derrick platforms are so devised that they can be carried up with the towers as they are set, and eventually will work from an elevation of 560 feet. In this position, the cable on each derrick-fall will be approximately 9,000 feet long -nearly two miles.

A shed is often built around it, and there he works the drums entirely by the clanging signal bells that a signalman away up on top operates. When the derrick is hoisting from the street, the signalman takes his signals from the men in the street, as he peers down from his eerie stand on some projection away up on the edge of the structure. When the derrick is setting, he moves in to a point near the base of the derrick and takes his signals from the derrick foreman.

As we know from our physics, the number of parts of line that pass back and forth through the separate sheaves of the pulley blocks determines the leverage or lifting power. The derrick itself must, of course, be strong enough to stand the greatest strain that will be imposed upon it, but the lifting power of the cable depends on the number of "parts" in the topping lift and fall. Now, when there is a very heavy load to lift, the number of parts in these two is increased, the blocks, of course, having suitable numbers of sheaves to do this. Roving up for a heavy load is a tedious operation, as the boom has to be laid down while the cables are passed over and back for the new roving. When not in use for heavy lifting, the fall is roved with fewer parts, for the increased lifting power is gained at the expense of speed. The fact that a three-quarter inch steel hoisting cable will stand a strain of over twenty tons before it will break does not tell the story. A lift of, say, fifteen tons or more must be gingerly handled. It would never do to start it with a jerk or attempt to hoist it at great speed. The shock on the equipment would be too great. The roving of a number of parts on the hoisting gear not only adds safety to the hoisting cables, but it gives the necessary slow motion to the heavy lift, a ticklish operation at best.

A brand new set of cables is almost always used at the outset of every large job, and the length of cable depends on whether any of this heavy lifting is to be encountered in the upper part of the structure. If there be such, the cable of the fall will be very long. For example, the fall cable for a building three hundred feet high, if we have to prepare for a lift requiring six parts to the fall when that height is reached, must be twenty-one hundred feet long, plus a safe amount wound on the drum, plus allowance for turns and indirect leads—perhaps a twenty-five hundred foot cable—nearly half a mile.

Courtesy of Thompson-Starrett Co.

Derricks in action—another view of setting the great trusses of the Paramount Building, New York.

CHAPTER XVI

SKYSCRAPER BUILDING–STONE

It may seem a far cry from geology to the skyscraper, and yet the two are in many ways intimately related. Some one has said that all building is founded on four elements; earth, stone, metal, and wood. Three of these are geological, and yet there is nothing in the science of building that requires the builder to be a geologist. Only love of the profession will prompt the builder to lift the corners of the pages of geology, for here is a vast storehouse of wonder that bids him look a little deeper, that the grandeur of the thing he does may take on a richer meaning. Yet we all know that, even in its profundity, the science of geology only scratches the surface of the mysteries it reveals. So it would be too much to say that a busy builder should be a geologist as well.

Our metals we receive after many processes have converted them into the useful parts we assemble. Our clays come to us after they have been fashioned into bricks and terra-cottas and tiles and sanitary wares. Our woods come fashioned from the mill after they have been kiln-dried, and even after they are in place, they must be filled and varnished and treated. The cementaceous rocks go through vast transformations before they come to us in the greatly changed form of Portland cement and the wall plasters.

But stone comes direct from nature to us, and its beauties are the natural beauties that nature has bestowed—no processing here except the process of hewing out of the solid the architectural forms, that the natural colors and textures may splendidly adorn our structures. Stone, the symbol of permanence, has from the very beginning been man's greatest and most everlasting building material.

Much of our beautiful marble comes from the Georgia Marble Company's quarries at Tate, Ga. The black appearance of the white stone is caused by the shadows.

Putting together or setting the great statue of Lincoln in the Lincoln Memorial, Washington, D. C.

Granite blocks from which great statues may be made. Clear, beautiful, white granite from the Mount Airy, N. C., quarry, ready to be removed to the cutting shed.

In the whole development of skyscrapers, our architects have never challenged the supremacy of stone; for while brick, that handmaiden of stone, has taken its proper place as a heritage from all that was best in the earlier forms, stone has been supreme as the basic material. Look about the great cities where every conceivable style and experiment of design may be seen; yet the stone base rules practically all. Occasionally one sees brick starting a foot or two from the sidewalk, but it gives the impression of instability, a sort of unfitness, no matter what the architectural attempt may have been.

And so we have the stone base reigning supreme, both because of a fundamental instinct that demands it, and for the entirely practical reason that it is in fact the best and most durable material with which to start the base of a skyscraper. Those few aristocrats of tall construction that boast a stone exterior to the very top hold proud distinction among their neighbors, and while not a whit more utilitarian, the distinction is there, and their owners take a pride in them that justifies their increased cost.

The exterior façade of metropolitan skyscrapers almost always starts off with a base course of granite. Early builders used to call this the water table, although modern design, with such strict limitations on extension over the building line, has forced the "table" feature almost into a mere rudiment, and we now commonly see the plane of the granite face on the building line, the same that is held for the general façade of the building; only cornices, occasional ornamentation, and the drips of sills extending beyond the line.

Geologists tell us that granite is igneous rock, which means that it was produced by those tremendous, pre-geologic heats supposed to have arisen in the molten condition of the earth when it was just emerging from its nebulous form. This same geological theory ascribes to the rock formations in Maine, New Hampshire and Vermont the earliest solidifications; but how-

Inside a modern granite cutting shed showing extensive tubing to serve the pneumatic hand tools. The large suction tubes carry away the dust and in a large measure protect the cutters, who are subject to silicosis.

Granite cutting by machinery is a recent development and much cutting is still done by hand. Its extreme hardness rendered it proof against machine working until the invention of tools of tremendous strength and power.

Courtesy of The John Swenson Granite Co.

ever that may be, practically all of the granite used in the cities east of the Mississippi River comes from the New England quarries or those located in the Appalachian Mountain range. Thus, we have excellent granite quarries in Georgia, competing even as far north as New York with the granites from New England; and while, throughout the South and Southeast, New England granites may be found, they are selected for the peculiar qualities and texture arising out of the architectural requirements rather than the structural requirements of the buildings in which they are used.

Architects select granites largely for their beauty and special effects of finish, because all building granites are practically of equal structural value, and as a matter of fact, all are much stronger than would be demanded by any structural requirements. The hardness and density of granite are the things that commend it structurally and are reasons for using it in the base course. It will withstand the most severe wear, but in addition, it conveys the feeling of solidity and everlasting permanence that commends it to the designers and owners of these great structures.

Unhappily, it has one quality which condemns it, and that is that under excessive heat it cracks and spawls inordinately, much more than almost any other stone, although we know that the stone work of any building, whether granite or other material, long exposed to great heat will be destroyed, either by spawling or, if it is of limestone origin, by calcination; that is, it turns to chalk.

Some of the granite quarries of the country are very old, almost as old as our colonial civilization, for throughout New England we can find tombstones dated as early as the beginning of the seventeenth century which can be recognized as having come from these quarries. Such stones, taken in the crude beginning from the overlying and easily accessible strata of the site of later quarries, indicate the peculiar nature of

Courtesy of Col. U. S. Grant, 3d, Director of Public Buildings and Public Parks of the National Capital.

Setting one of the huge Colorado Yule drums of the colonnade on the Lincoln Memorial, Washington, D. C. These stones weigh nearly twenty tons each.

Courtesy of The John Swenson Granite Co.

A well-opened granite quarry at Concord, N. H. As the quarry deepens the strata become thicker, the result of the geological cooling processes. The lower layers, cooling more slowly and under tremendous pressure from the over-burden, solidify with less frequent stratifications.

granite in its formation. Almost any granite quarry, when the
overburden of earth has been removed, reveals on its top
surface thin layers, anywhere from a few inches to a foot
thick, actually separated from each other, so that, with a little
skill, those early quarrymen were able to break off rough
quarry sizes for their small tombstones. As the quarrymen go
deeper into the quarry, these laminations become thicker, and
sometimes ten or twelve feet below the original surface, the
better and thicker stone, suitable for modern building, is
found. A quarry thus far developed is apt to be pretty well
established, and the experienced quarrymen know that the
color and quality of the stone from there down to almost unlimited depth is practically assured—or if inferior quality is
encountered, it will cause abandonment of further working at
that point.

In the great quarries at Barré, Vermont, and throughout
New England one sees, running away from the quarry, small
industrial railroads that have through years of quarrying built
up vast piles, almost mountains, of quarry tailings; almost all
the waste of these layers where they thin out at the edges.

Formerly, the splitting off of dimension stones from these
quarries was carried on by first drilling lines of holes about an
inch or so in diameter to a depth of perhaps two or three feet.
This was called line drilling, and the splitting was done by
means of "plugs" and "feathers," a method still in use in
some granite quarries. A feather is nothing more than a half-
round piece of iron bent away from its flat side at one end.
Two of these placed with their flat sides together will present
the appearance of a small, round bar, about six inches long, that
has been split in two for its entire length. Two feathers are
put in each hole, the joints all in the plane of the line of holes,
ready for an attack by the plugs. The plugs are nothing but
iron wedges which are inserted between the spread ends of
the feathers. These are first lightly tapped into place, and as

soon as they are secured, men walk back and forth along the lines of plugs driving them in, thus exerting in the aggregate an irresistible force, which soon splits the whole block of granite along the line of holes. The bottom of the laminated layer lies somewhere below the sheet on which the quarrymen are working, and the quarry foreman knows about where this is. If the work is being carried on in a place where the easiest quarrying occurs, these laminations will be about four to six feet thick. Some quarries have hydraulic pressure equipment, which forces water under tremendous pressure into pipes drilled six to ten feet into the rock, and sealed where they enter the rock, so that the pressure is exerted at the depth of the pipe in the rock. Such pressure will split the granite as effectively as will line driving.

The large blocks, split off by either method, are again broken to derrick sizes, and a derrick, conveniently placed, raises them and sets them on cars, to be carried to the cutting sheds. The cutting sheds may be located either near the quarry or at some distant point.

Granite, on account of its hardness, cannot be worked by power tools as easily as can some of the softer stones, such as limestone or even marble. However, the invention of automatic tools and tough, alloy steels has opened up a field of machine working in granite of which the cutters have not been slow to avail themselves. The cutting shed of to-day is a place of ingenious, automatic tools, guided by the same skilful hands that, only twenty years ago, laboriously pounded away at the granite surfaces to shape them into the architectural forms of the buildings of that day.

Like all other building-stone, granite is cut to cutting diagrams with due regard to its fitting the structural steel, and its anchoring to the steel frame of the building. When it arrives at the building, there still may be a little fitting to be done, so we see hovering around the setting of granite both the setters and the fitters—two separate trades, an example of the high division of building labor. It is true that on many jobs there is work for both setters and fitters, but frequently the fitter has many idle hours on his hands as he watches the perfectly detailed granite set in place. Granite, like other forms of exterior stone, has to be anchored to the steel, and the fitter plays his useful part in fitting around beams and cutting special anchor holes in the edges of the stone. Frequently a forge is near at hand to facilitate the bending and twisting of these anchors better to secure the stone to the steel frame; and when the stone has thus been anchored, it is backed up with brick work.

Granite, and in fact all exterior stone work, is only a facing from four to eight inches or more thick, never solid through the walls of a skeleton steel structure, first because it would be too expensive, and also because brick backing provides the proper foundation for the interior finish.

The stone is hoisted into place by derrick men, still another trade, useful and busy; but adding another to this group of trades around the stone setting operation, which includes the fitter, the setter, the derrick man, the bricklayer, and occasionally the blacksmith, while several of these trades have their laborer helpers. Yet, people sometimes wonder why stone work is so excessively expensive.

The granite facing of a building may extend to any height, from the mere base course projecting a few inches above the sidewalk level, to the last stone on the cornice of the building. Due to the high cost of granite and the expense of setting and fitting, we see fewer and fewer all granite skyscrapers. Nevertheless, there are some beautiful examples of this extensive use of granite. The Wanamaker store in Philadelphia is granite clear to the cornice. The Hanover Bank building in New York, twenty stories high, is similarly an all-granite façade. In both cases, the method described for producing and setting the granite was the same. These are, no doubt, everlasting structures, unless, perchance, a destructive fire overtakes them, either from the burning of their contents, or fires occurring in adjoining buildings. If such misfortune ever overtakes them, the beautiful façades will most certainly be sad sights after the spawling by conflagration, aggravated by the added havoc that the water from the fire hose playing on the hot stone would produce.

Every farmer knows that the way to get rid of a granite boulder in his field is to build a fire alongside of it, and when the stone becomes hot, to throw water on it. It bursts asunder under the swift contraction that the drenching of cold water causes. In like manner, the granite front of a building will go to pieces under such alternate intense heat and cold. In the Baltimore fire, several beautiful buildings with granite façades seemed to suffer the most from the fire. The excessive damage was due to this weakness of the granite.

The high cost of granite and its difficulty of working are in part responsible for its limited use, but not altogether. Architects have found that the combination of granite and some other stone makes a pleasing and dignified façade.

The most commonly used building-stone, which in the eyes of many is also the most beautiful, is limestone. It can be recognized by its warm, grayish or buff color and sand-like texture. It is in many respects an almost ideal building material on account of the ease with which it is worked and the beautiful effects that are obtained by the use of it, both in its plain surfaces and where moulding and carving are desired. The reader must not get the impression that limestone, wherever found, is suitable building material. In fact, fine limestones, so commonly observed in metropolitan structures, are found only in a few parts of the country. The most extensively developed quarries are those in the neighborhood of Bedford, Indiana; hence the term, Bedford stone, which is loosely applied to the particular form of oolitic limestone generally used as building material.

Limestones are of wholly different origin from granite. Geologists tell us that they are of sedimentary origin; that is, laid down by the deposit of waters from the eroded and dissolved calcareous elements that the rivers brought down to the geological seas and lakes. Again we are in the field of profound speculation, because in developing a theory of the origin of this material, the geologists are uncertain as to whether all of that limestone came from these erosive and sedimentary phenomena, or whether it all came from the organic skeletons deposited through geological eras in infinite numbers in the beds of those old geological seas.

Marble, it is to be said, is of the same general stuff, all originating from the calcareous formations of the earth, a stupendous thing to contemplate.

Limestone quarrying is much simpler than granite quarrying. The stone is softer and yields easily to the quarrying equipment, which is now almost universally power driven. Like granite, limestone is generally quarried from the surface after stripping away the soil and other over-burden. The oolitic stratum, when reached, is easily recognized, and on the exposed surface after stripping, the power-driven channeling machines are set up. Long channels are run, and in these wedges are inserted. Limestone, however, is not laminated, and a quarry, to be successful, must have from twenty to seventy-five feet of thickness to justify working it. This oölitic deposit is seldom found more than a hundred feet thick.

Blocks of limestone having been broken off like granite, they are transported to cutting sheds; but unlike granite, limestone yields almost wholly to power machinery. Its relatively soft texture and homogeneity are such that it can be safely worked with power tools. Thus, the limestone cutting shed is more like a modern industrial plant. The old stone cutter of

Limestone quarry. This sedimentary deposit laid down in the geologic eras when seas covered the central part of the United States. Note that the oölitic limestone is sharply divided from the over-burden, which is called Mitchell.

Channeling-machines blocking out the limestone. The huge monolith being pulled out is too large to be lifted intact and is cut up into what is called quarry stock. The limit of lift and handling in transportation is about 125 tons.

Courtesy of Indiana Limestone Co.

the maul and chisel has been almost wholly replaced by power machinery; yet this machinery is supervised by stone cutters, and it still remains that the last finishing touches of fine carving or the joining of delicate mouldings has to be done by hand.

The setting of all sorts of exterior stone work is done by the same method; the same tools, the same workmen, the same equipment. However, the brickwork backing of limestone is generally laid in non-staining cement, for ordinary Portland cement has the property of discoloring limestone. Many devices are used to prevent this, one of them being the painting of the stones on the back and edges with some material, generally an asphaltic paint which is intended to seal the pores of the stone against the discoloring ingredients of the Portland cement. Non-staining cement is especially prepared with these damaging ingredients eliminated.

Limestone seems to have established its place forever in the hearts of designers, and is the accepted badge of dignified gentility where used to any considerable extent for architectural effect. We commonly find limestone reaching to the second or third floor of the skyscraper, and at this point it is often terminated with a handsomely moulded cornice or frieze. The windows which penetrate it are enhanced and beautified by its mild but virile texture, and ornamentation carved in it, if well designed, is sure to be striking in its clearness—a most appropriate architectural medium.

After the cornice on the second or third floor, the building is apt to rise a sheer brick shaft with only sills and possible lintels of limestone. Then come the main frieze and cornice, which, if not of limestone, will probably be of terra cotta in imitation thereof; for now the appeal to appearance is farther removed from the eye, and mass is apt to take the place of beauty of detail and texture. Not infrequently, the shaft of the building is made of limestone, one of the richest of exteriors,

Quarry stock stored ready for transportation to cutting sheds in all parts of the country.

Columns for a mighty edifice ready for transportation.

The turning-lathe adapted to stone cutting.

and in the hands of a skilful architect, it is one of the loveliest materials to be found anywhere.

Marble is a near relative of limestone geologically. In certain parts of the country where marble is quarried, particularly in Tennessee and Georgia where very pure strains of marble occur, the waste is burned into lime in adjacent kilns, —one of the sources of the ordinary lump lime of commerce, with which builders have been familiar from time immemorial. But it is of marble as a building exterior that we are speaking, wide in its variations of color, texture and figure. Not so frequently do we see figured marbles in exteriors. Architects generally prefer the natural shades without variegations, sometimes streaked with saffron or tan, sometimes with blues and grays, slight color variations that give the marble warmth, as artists would say. As a near relative of limestone, it is supposed to have a somewhat similar origin, but it is of vastly greater antiquity. Moreover, it is thought that, succeeding its limestone origin, it was transformed into the marble we know by tremendous pressures, due to the contraction of the earth's surface; those pressures which threw up the mountain ranges. For marble is seldom found in horizontal sedimentary beds, the strata generally indicating clearly the original sedimentary origin, afterwards disturbed by the titanic upheavals that have brought up the out-croppings we see in marble quarries. Heat, too, may have played its important part in the formation; but of this we shall not attempt to speak, beyond noting that, out of this cosmic evolution, with its unnumbered changes and variations, the beautiful marbles are a product of the admixture of original, basic limestone and mineral colorations, of which iron, cobalt, feldspar, and an endless variety of other ingredients play their important parts. Marble is a carbonate of lime; so is limestone; so is onyx; so are the dolomites. The unlimited variations may be appreciated.

Marble, like limestone, is worked almost wholly by ma-

Left—Gang-saws sawing a large block of Indiana limestone into commercial sizes. Varying thicknesses are obtained by changing the spacing of the saws. *Right*—This diamond-studded saw does the work of 100 men.

Left—Inside a cutting-shed at Bedford, Ind. Limestone is worked almost entirely by machinery, the finishing work being done with hand pneumatic tools. The finest finishing touches are still done at times by expert artisans with hand tools. *Right*—Cutting the voussoir of a beautiful Gothic arch, an example of the triumph of mechanical stone cutting.

Courtesy of Indiana Limestone Co.

chinery. The blocks, quarried after the manner of limestone, are sent to the cutting sheds. We are now speaking of its use as exterior ashlar, where the blocks are cut in thicknesses ranging from four to eight inches, some less, and of course, to greater thicknesses where cornices or thick and heavy members are required. In the cutting sheds, it goes through power-driven gang-saws to be reduced to ashlar. These are not unlike wood saws, excepting that they have no teeth. They are rapidly driven back and forth by machinery, with quantities of water and sand pouring over them as they work. The blocks of ashlar thus produced pass on to plane beds or turning lathes, according to the requirements of the cutting diagrams, the same system of cutting diagrams as is used for any sort of exterior stonework.

Marble is more of an aristocrat than limestone. It finds its way into our most beautiful public buildings and carries with it the seal of aristocracy. But it can hardly claim greater dignity than limestone. It is much more expensive, both on account of its comparative rarity and on account of the added difficulty of cutting, because, while cut on the same sort of machine, the cutting is harder and the processes are therefore slower. We know, of course, that marble is susceptible of much finer and more beautiful carving than is limestone, or in fact, any other stone. In its finest grades we think of the beautiful Carrara marble statues, the medium used by the Greeks and Romans, and in fact, through all ages where the finest stone statuary has been produced. The Elgin marbles from the Parthenon, the greatest classic of architectural carving of all time, themselves proclaim the perfection of that stone as a medium for the most beautiful building sculptures we know.

There are varieties of marble, beautiful in color but unsuitable for exterior use; but these we shall mention later when we are discussing interior finish. In the warmer climates where the ravages of alternating seasons of frost and heat are not to be contended with, some of these colored building marbles are most effectively used. And here other relatives of limestone are used with great effect. The principal one is travertine.

Now travertine is a generic division of the limestone family which bears an evil reputation on account of its friable nature. Geologists sometimes call it tufa, and it is generally dismissed by them as having no commercial value. Far different is the Italian or Roman travertine. From these ancient quarries on the banks of the Tiber was cut the stone for some of the noblest of the Roman exteriors in the heyday of Roman supremacy in architecture. The supremacy of the travertine is proclaimed by these magnificent Roman specimens that still survive.

Travertine is porous, very porous; in fact, close inspection reveals that the stone is riddled with small interstices and fissures, but still, it is a very solid stone and at a short distance presents a most beautiful appearance. It is generally light buff in color, although there are darker varieties of it. In America it is being used increasingly for interior finish, but very little for exteriors because of its porous nature, for in spite of its artistic appearance, the interstices will fill with water and ice in our northern climates, the forerunner of a general breakdown of any water-holding surface exposed to this rigor.

There are, of course, endless varieties of other types of stone —sand-stones, dolomites, and other branches of the limestone family. In different parts of the country some of these are favorites on account of their availability and beauty. There is no intention here to attempt to list and classify the wide varieties of building-stones available; rather, it is intended to point out a few of the most conspicuous, with the idea that the reader's interest will be attracted to inquire as to just what the stone is, as he sees a skyscraper in course of construction.

Stonework, of whatever sort, is said to be jointed when there is careful cutting of the surfaces that come adjacent to each other to form uniform joints, as distinguished from the haphazard joints that occur in rubble stone masonry, where the irregular sizes and shapes are laid with mortar joints of rough and uneven bedding—anything to insure a solid wall with the beds and ends supplemented by spawls and chips. Such rubble is a useful and permanent form of masonry in its place, but is seldom seen in the exterior of a skyscraper. Almost universally, the stonework of the tall metropolitan structure is carefully cut and jointed. The horizontal joints are called bed joints and, as their name implies, form the beds on which the succeeding courses are laid. The end joints are end or vertical joints, and the joints in arches where they occur are called radial joints.

Generally all of this jointing is carefully laid out on the architectural drawings In fact, the jointing is the subject of much study on the part of the architect, and reflects one of the oldest of the sciences connected with building. Stereotomy is the science of stone jointing, and in the days of pre-skyscraper construction, it held its structural importance even as the science of structural steel design now controls the structure of a modern skyscraper. Stereotomy reached its highest development in the Renaissance, and perhaps was the only true science that entered into the construction of the great cathedrals of Italy, France and England. The architects of that period developed the science to an amazing degree, as the mighty cathedrals so amply testify; but in England the development was carried to perhaps its most advanced form, and indeed it is probable that the science was completely understood and its absolute limits attained in some of the great English structures. The fan vaultings at both Cambridge and Oxford, not to mention the stonework in the imposing cathedrals of Winchester, Canterbury and others, are still the marvels of the

world in the intricacies and studied ingenuities of their stereotomy. In some of our best architectural text-books are to be
found scale and isometric drawings of these wonderful creations of the stone-cutter's art. Stereotomy is still taught in
many of our engineering schools in connection with descriptive geometry.

We think of stone as the everlasting material, which it is;
but stonework, when its laying and jointing are considered, is
not everlasting. The fact is that the jointing of stonework,
particularly where copings and horizontal surfaces are concerned, has a bearing on permanence, for the work must be so
designed as to admit of the least possible penetration of water.
Especially is this true in our northern climate where the action of alternating frost and heat must be contended with.

Experience shows that all finished stonework, and in fact
stone structures of almost any sort, must be watched and kept
up through frequent inspections, and occasional rejointing
and even resetting of copings, platforms and balustrades. The
casual observer does not see this, for it is all done unobtrusively, but it is done nevertheless. The perfection of the stonework
of the Capitol at Washington, to mention only one of our
great monumental stone buildings, gives the impression of
everlasting permanence, but this is gained by constant vigilance of upkeep. Let any such structure go ten years without
this care and it will commence to show very definite signs of
a start on the road to dilapidation. The trouble comes not
from the stone but from the exposed joints, which yield under
the action of expansion and contraction—alternate heat and
cold.

The sedimentary stones, limestone, marble, sandstone and
the like, seem to be most enduring when laid on their natural
bed, that is, where the bed joints are cut parallel to the horizontal plane in which the stone was originally deposited by
nature. This is not always easy to do, because the finer grades

of these stones are so homogeneous as to make it almost impossible to recognize the position of the natural bed after the quarry blocks have been removed and perhaps turned over a few times in the handling. Some of the sandstones almost demand natural bed setting; and if the cutters fail to observe this, the result in a few years is a noticeable deterioration. Witness the scaling surfaces of many of the old brownstone fronts of New York and other cities. Where this scaling is most aggravated, it is a sure sign that the natural bed has been turned on edge and laid parallel to the plane of the wall, thus exposing the laminations to the elements.

The finish of stonework may be anything from rock faced —where the surface, as its name implies, is simply roughly split from quarry sizes—to honed or even polished surfaces. Only the hard stones, such as granite, certain marbles and a few dolomites, can be polished; the soft stones, such as limestone, the soft marbles and sandstones, cannot be given a higher finish than rubbing or honing. Architects have always known the technic of finish, and some very beautiful effects are obtained by the mere change in texture of the cutting. Thus, in granite or other hard stones, the surface may be said to be six-cut, eight-cut, or even ten-cut; meaning that the bush hammers used to produce the finished surface have six, eight or ten blades to the inch.

Soft stone surface dressing is effected with maul and chisel, the chisel having a comb-like cutting edge which the artisan skilfully guides where the six, eight or ten-cut effect is desired. This refers to hand work on carving and irregular surfaces. On plain surfaces and straight mouldings, the effect is produced by shaping the planing tool on the machine, for soft stone and indeed marble are almost universally cut by machinery. The bed rubbing or honing is generally obtained by laying the stone face down on a large turning table that whirls horizontally, the attendant simply seeing that plenty of sand

and water are at all times kept in the path of the stone as the turning disc abrades the surface.

There are, of course, a great many ways of cutting stone surfaces; the few here mentioned are given simply to lead the reader's attention to some of the standard cuttings, so that the effects of surface appearances will be better understood.

There are notable buildings of granite that are not of sky-scraper type, but which demonstrate the perfection to which the granite-cutter's art may be carried. The State, War and Navy and the Treasury buildings in Washington are examples of beautiful granite cutting, whatever one may think of their architectural design.

Probably the greatest stone cutting and setting job of all time, where perfection of workmanship predominated, the number and size of massive stones considered, is the Lincoln Memorial in Washington. It is of Colorado Yule marble with Indiana limestone interior walls and pink Tennessee marble floor. The statue of Lincoln is of Georgia marble. Again we have in this building some of the principles of the skyscraper, for structural steel is used extensively. Henry Bacon, the architect of that superb structure, who gave the last fifteen years of his life to the study, design and erection of the Memorial, is quoted as saying that it contained more large stones than any other structure ever built. Each stone in the stylobate, as well as the individual stones in the drums of the columns forming the great colonnade around the building, weighs over twelve tons, —most of them from sixteen to eighteen tons each. The Indiana limestone lintels of the interior weigh over twenty-five tons each, and in the whole structure there are over a thousand individual stones ranging in weight from twelve to twenty-five tons.

Here in this structure, stone cutting and stone setting probably reached the highest perfection ever attained in American construction, and it is doubtful whether it will reach these heights again for generations to come. Its mighty dignity and impressive beauty could only have arisen from the magnificent human attainment that the Memorial was built to commemorate.

Courtesy of Kittanning Brick & Fire Clay Co.

Interior brick-making crew showing off-bearing belt, pug-mills, automatic cutters, drying-cars, etc.

CHAPTER XVII

BRICK, TERRA-COTTA AND THE CERAMICS—THE TECHNIC OF CERAMIC MASONRY

THERE is always friendly contest between the advocates of various building materials as to their durability and usefulness, and having discussed stone as a building material, consideration of its great rival, brick, should come next in order, for brick has always been a worthy competitor of stone. Of late years, the archeologists have actually elevated the prosaic matter of brick work to the dignity of a news item, owing to their discoveries of the ancient cylinders on which the cuneiform records of the Chaldeans and earlier peoples were kept. The ancient city of Ur has almost been reconstructed in our minds by the unearthing of great mounds of brick which turn out to have been their libraries of public and private records. Those ancient civilizations had learned the art of burning brick, the earliest excursion in the field of ceramics, yet they ran side by side with another civilization which had not advanced so far, but rather had constructed its brick merely by drying clay, mixing it with straw as a binder. Thus, the biblical reference to making bricks without straw reflects an ancient lament based on a really practical need. It was impossible for those old Hebraic brickmakers to carry on their art without straw, and they let the world know about it.

But it is of the burned brick we speak, because that art seems to have been known to all civilizations. The Orientals had it, as did the races of the Near East. It is not the intention here to trace the history, but rather to show that in all civilization burned brick has played, and still plays, an indispensable part. Modern sporadic excursions into the field of Portland

ENTRANCE DETAIL.

Theodore A. Meyer, Architect.

A carefully prepared brick lay-out, showing courses and jointing, and its appearance after construction.

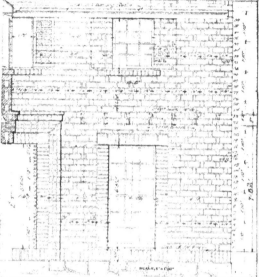

Courtesy of The Architectural Forum.

Flemish bond emphasized by black headers.

English bond—alternate courses of headers and stretchers. Two patterns are
here revealed by the use of dark headers and stretchers.

Courtesy of Fredenburg & Lounsbury.

English bond that is made to look like Flemish bond through the use of dark and light brick.

An interesting mixture of Flemish and English bonds.

Running or stretcher bond, sometimes called American bond, with headers every fourth course; an adaptation of the English bond. Note the color variations and deep joints to give color effect.

Courtesy of Fredenburg & Lounsbury.

cement and pressed lime brick have in no way disturbed the supremacy of burned brick, and to-day it should not be thought of so much as a rival of stone but as a complement and necessary team-mate of stone in the production of our sky-scraper. The possibilities of its color variations are legion, from the very whitest to black, ranging through all tones and styles. Modern brick makers can produce almost any color, yet certain fundamentals seem to hold their place even though styles and ornamentations change.

It is not necessary to describe the mere act of bricklaying; that is too commonplace and every one knows how it goes. Yet the building of the curtain wall of the modern skyscraper calls for a technic that may be summed up in the term "everlasting vigilance." These great, exposed wall surfaces of sky-scrapers receive from the elements a punishment that is little appreciated by those who sit warmly behind the protection they afford. Yet, in the production of the walls so tight and safe, the builder's art has been exercised to its utmost. First the wall must surround and completely encase the exterior of the structure. While it is carried from floor to floor on spandrel beams, the wall must be so surely and truly laid that the driving storms will not penetrate. When water drives through, as it does all too frequently, the brick work is at fault. Therefore, the competent builder exercises great care in seeing that the work is solidly built, solidly tied in to the steel and carefully pointed on the exterior. The whole object is to produce an impervious exterior with the joints between the windows and brick work thoroughly caulked and sealed.

And having touched upon this prosy technic, we are back to the interest in color and appearance, for the visual appeal is the important thing to the observer, even if it is not to the user of the building. Perhaps the first thing observed is the "bond" of the brick; that is, the manner in which the outside courses of brick are bonded to the body of the wall. There are in fact

Two interesting variations of Flemish bond to give "texture."
Texture may be further emphasized by deep or flush joints of various kinds, and mortar colors.

Courtesy of Fredenburg & Lounsbury.

only two real bonds—the so-called English bond, where a row of "headers" or ends of the brick are alternated with a row of "stretchers," the sides or long dimension of the brick. The other is the Flemish bond, where a header and a stretcher are alternately used. Where walls eight inches or thicker are built —and brick walls in skyscrapers are almost universally about twelve inches thick—it is apparent that brick thus laid will knit in and join with the body of the wall, and when thorough work is done, the result is a homogeneous mass and, as has been said, impervious.

Behind the face brick, which are generally a better and more selected quality, comes the backing of common brick. Common brick are inferior to face brick only because they are common. They are just as well burned, just as hard and just as durable a building material. Face brick, as its name implies, is simply an exterior color dressing to add an artistic effect to the solidity of the wall of which it is a part.

The bonds of a brick wall are not to be confused with its texture or color effects obtained by mixing different colors of brick, making endless variations of the two basic bonds, or of laying brick in unusual ways. Thus, a brick may sometimes be laid with its end exposed, but standing on edge. Bricks thus laid are technically known as rowlocks. A rowlock course is frequently shown, and called for in specifications. Similarly, when brick are laid vertically—their long dimension in a vertical position—they are spoken of as soldier brick; a soldier course is here called for. With these variations of position and variations of color or shade, with variations of bond spacing, an infinite variety of pattern is obtained in the façades of the buildings. Some of our most beautiful effects of textures are nothing more than arrangements of brick jointing and shading, all variations on the play of the English and Flemish bonds.

Occasionally one sees a continuous, running bond; that is,

without any headers showing. It has its uses in small areas and for certain artistic effects, but it always brings a shudder to the eye of the trained builder. It produces an unnatural effect, because the builder's eye demands an evidence of the face being bonded to the body of the wall. Where this effect is desired, it is necessary to tie the face to the body of the wall with metal clips, a questionable practice; or every few courses, the corners of the face brick are clipped off, and common brick running diagonally into the body of the wall are inserted into the interstices left by these adjoining clipped ends. This is better than using metal clips, but is not as good as the substantial and natural bonding by the old-fashioned use of headers and stretchers.

Face brick used to be called pressed brick a few decades ago, and the American fashion of bricklaying during the period of our architectural decadence—say from the Civil War until the World's Fair—demanded great precision and accuracy of brick sizes and bonds. Those were the days of the old, red brick fronts, for during that period there was not available a great variety of texture and color. A red brick front was the symbol of stolid elegance.

It was not until skyscrapers had come well into their own that architects commenced realizing the possibilities of texture and color variation. Suddenly they discovered that beautiful effects were to be obtained by mixing the shades of brick; that precision of brick sizes and laying produced too formal a result. In this new field of variegation and variety of texture, the brick makers took a leading part, with the result that today the warm and pleasing effects of our best façades are largely obtained by a freedom from precise colors and mechanical perfection of workmanship.

In seeking variations in color and warmth of texture, many new ways of cutting and moulding the brick were devised. One large brick manufacturer got the inspiration of calling his product "tapestry brick." The name seized the imagination of the public, which at once commenced to apply it to all warm, variegated brick surfaces of interesting texture, particularly if tans and browns prevailed. Strictly speaking, tapestry brick is the proprietary product of a prominent brick manufacturer, but the public, no doubt, will continue appreciatively to designate its favorite brick surfaces as tapestry, regardless of the manufacturing origin.

We have said that common brick is just as good and just as hard as any other brick. This, of course, means good common brick, for in the market place we hear of "salmon" brick and "clinker" brick. The former is admittedly inferior, is more porous, less durable, and perhaps even unsafe to use Salmon brick comes from the part of the kiln farthest from the fire. The ancient brick makers expected to obtain only a certain percentage of good, hard brick out of any kiln. Salmons were the necessary and unwelcome by-product, which were sold at any price they would bring and without recourse.

From the other side of the kiln, that is, the part nearest the fire, come the clinker brick. These are exposed to too great heat and show a tendency to melt and become deformed under the weight of the bricks piled upon them in the kiln.

In their search for variegation and variety of texture, architects occasionally hit upon clinker bricks for certain special effects. They are rough, irregular, have a tendency to bluish-gray color, and are almost vitreous in their hardness. We do not see clinker brick used in large, metropolitan buildings as frequently as in country residences, but wherever they are used, they are durable and sound, however unsightly the individual brick may appear Some one has said that beauty is in the eye of the beholder, an aphorism that is fully proven where extensive use is made of clinker brick.

The process of brick manufacture, like other materials entering into building construction, has undergone an improvement that has kept pace with the development of the building industry itself. For thousands of years brick were produced in the same old way. First the bricks were cast in moulds. Then they were dried, or partially dried, under protection from the rain and the elements. After the drying process, they were stacked in kilns so as to admit of fire being built in tunnels or flues running through the kilns. Wood, and later coal, was used as fuel, and after several days of burning under continuous heat, the fires were allowed to die out, and then for several days more the thoroughly heated kiln was allowed to cool. As soon as the bricks were cool enough to be handled, the kilns were torn apart and the bricks assorted according to quality, the salmons, as has been noted, cast aside as inferior, and in earlier days, the clinkers too were thrown out as utterly worthless. The main body of the burning was the net product of sound brick, and the brick maker's art turned upon his ability to produce the fewest salmons and clinkers, and of course, the greatest number of sound brick.

Under any of two or three modern processes, brick making is a very certain operation, and the waste in clinkers and salmons is relatively small. Nowadays, the moulded brick from the brick machines are piled on small cars, shoved into long drying tunnels, where the heat coming from the kilns passes over the brick, thus greatly accelerating the drying process. The arrangement is such that the cars pass very slowly through these drying tunnels, which are so constructed as to admit of the greatest possible length of trackage back and forth as the cars approach the kiln. As they approach, the heat grows more intense, and by the time the cars have passed through the full length of the tunnels, the moulded brick are dry and ready for stacking. From here they pass into permanent kilns scientifically arranged, with the firing so devised that the heat is almost uniform throughout the kiln. The brick come from the kiln without the necessity of tearing it down. These kilns are sometimes called "Dutch Ovens," although there are many modern forms to which this could not be applied. Firing may be by coal or natural gas, and more recently, by crude oil.

The greatest advance in brick burning, naturally, is found in the manufacture of face brick, where the quality of the product warrants every refinement of manufacture, burning and handling.

When skyscrapers were first built, there was a problem in connection with laying the exterior walls. In the days before skyscrapers, scaffoldings could be constructed from the ground and carried up even to a height of six stories, as this was about the ultimate requirement. It is true that smokestacks of great height had been built before that time, and the thing was accomplished by laying the work from the inside, a scaffold and hoist being constructed all the way up. Working from the inside of a wall in the manner pursued in building stacks is called laying the brick "overhand." It is not an easy thing to do, particularly when fine workmanship is required. Even now we sometimes see the rear walls of buildings laid overhand, but to do fine face brick, it is absolutely necessary for the workmen to stand out in front of their work in order that the jointing and coursing may be truly and accurately done.

The early builders devised a system of outriggers built on the floors, which were in effect little cantilever bridges made on plank stood on edge, suitably trussed and braced, on which platforms would be built, and as the work rose, horse scaffolding would be constructed to take care of about every five or six feet in height. A story having been finished in this way, the scaffolding gang would go on ahead of the bricklayers as soon as the sills on the floor above had been laid, stick their "pudlocks" through the window frames and prepare another scaffold of the same sort on the floor above. In this way the process would be repeated over and over again until the top of the building was reached. The system presented difficulties espe-

Courtesy of Patent Scaffolding Co.

Topping out a building. The suspending cables have been wound up on the drums from the bottom of the building, and the masonry will be finished either from horse scaffolds standing on top of outrigger beams, or will be laid overhand from the inside.

The modern bricklayer in winter within a scaffolding enclosure. The salamander in the background keeps him warm, and his mortar is tempered in hot water.

Courtesy of Turner Construction Co.

NEW JERSEY TELEPHONE BUILDING, NEWARK, N. J.

Curtain walls and enclosed scaffolds, showing the complete dependence of the exterior walls on the spandrel beams of the steel frame. Canvas enclosures are placed around the scaffolding for the protection of the workmen in winter.

cially at the corners of the building where the pudlocks would have to stand out on a diagonal in order to carry the scaffolding around the corners. Thus, holes had to be left in the corners, which were afterward filled up by workmen being let down over the outside of the building on swinging scaffolds. On some of our structures built during that period, an experienced builder can still see the places where these corner pudlock holes occurred. At best the system was awkward and clumsy.

This method has been superseded by the hanging scaffold, which is now almost universally used. It is an ingenious device which, by means of little winches located along the scaffold, enables the workmen constantly to be working at the most convenient height, about waist high. As the wall is built up, laborers work back and forth along these scaffolds, taking up on the small winches, so that the brick work goes on continuously and without the interruption that the old pudlock system inevitably required. On very high buildings, it is sometimes expedient to set the outriggers in two sections, the first from, say, about the twentieth floor, while they are yet setting steel on floors away above this point. This enables the builder to build a great deal of the exterior brick wall before the steel is finally completed, thus saving time—the ever-present requirement of skyscraper construction.

Architectural terra-cotta, like brick, is a heritage from ancient times, and while it cannot claim the same early origin, we know that the art was highly advanced during the Renaissance, and to-day the work of the great Italian artist, Della Robbia, bespeaks the highest development of the ceramic art. Modern terra-cotta is actually somewhat different from the product of Della Robbia, but our modern manufacturers look to him as their god and their progenitor.

It is probably true that the modern skyscraper is almost solely responsible for the high development of the art of terra-

The fire which destroyed the wooden scaffolds surrounding the Sherry-Netherland Hotel, New York. The structure was little damaged by this fire, due to its fireproof construction.

A scaffolding frame made entirely of pipes with fireproof wood planking from which the ornamented copper roofing and bronze lantern of the New York Life Insurance Company Building was built.

Cathedral of St. John the Divine, New York, where light, modern pipe scaffolding supplants the cumbersome wooden work formerly used.

Courtesy of Chesebro, Whitman Co., Inc.

cotta manufacture. First used in our post Civil War decadent architecture as a reddish material, by-product of the old pressed brick works, it has, in the last twenty-five years, been raised to one of the most scientific of the liberal arts. The beautiful effects obtainable in terra-cotta so far transcend the dreams of its advocates of even twenty years ago, that in its modern form terra-cotta might be said to be a new building material.

We have spoken of it as a substitute for stone, because in one of its many variations, it can be made in almost perfect imitation of any of the building stones used in exteriors. Then, it is used as a material by itself in an infinite variety of shades and colors. Finally, as a polychrome, it is being used more and more, and the striking enamel finishes obtainable in this beautiful material have opened up new fields of architectural ornamentation.

Terra-cotta basically is made as brick is made. Like some of the higher grade face brick, it receives a surface coloration apart from the body of its structure. This surface finish is technically known as the "slip." In the development of textures and colors, ceramic scientists of the terra-cotta factories have drawn from all other branches of the science of ceramics in the production of colors and enamels.

Like stone work, architectural terra-cotta is made from diagrams, but owing to its plastic origin, it can be worked into a greater variety of shapes and forms. Moreover, terra-cotta is relieved of excessive weight by hollowing out the backs of the pieces, for it is to be remembered that it is seldom used structurally; and as a form of ornamentation, it need only be of sufficient strength to retain its position in the wall, and more essentially, to withstand the action of the elements. Good terra-cotta is everlasting, just as good brick is everlasting. In its inferior grades, the slip may be its undoing, for, unless the surface is carefully made and the firing done with skill and understanding, water will penetrate this material and the succeeding

Courtesy of Atlantic Terra Cotta Co.
The modelling shop of a modern terra-cotta plant. On the walls are models made for the architects.

Courtesy of Federal Terra Cotta Co.
Terra-cotta clay being pressed in the mould by hand, where it is left for two days before being turned out and dried. The mould was made from a plaster model.

Setting the finished body in the kilns after receiving the ceramic coating.

frosts will be its ruination. But it is of the better grade of terra-
cotta that we speak here.

One of the reasons for good terra-cotta is the everlasting vigilance of its manufacturers. The clay is carefully selected and
analyzed, even more carefully prepared, and throughout the
process, every step is watched; but before there can be any
making, the same vigilance must preside over the preliminary
work. We have seen that it starts with diagrams, a science in
itself, for the diagram draftsman must be cognizant of the
technic. All jointing must be carefully considered and anchoring provided for; the limitations of the material must be fully
understood.

Terra-cotta makers work with a special rule on which the
inches and feet are about eight per cent longer than the ordinary rule; thus, a ceramic worker's foot is about thirteen inches
in this shrinkage scale, and the fractional divisions are, of
course, proportionately extended; for it is to allow for shrinkage of the material in firing that this allowance is made.

After the diagraming, separate moulds are made for every
different piece. If straight-away, plain work, these moulds are
of plaster of paris; if heavily ornamented, they are made of
some form of toughened gelatine or a composition of glue and
plaster of paris—something to form a tough, durable mould.

Where elaborate ornamentation is involved, skilled modellers prepare clay models of the parts, and architect's specifications generally lay great stress on the question as to who the
modeller shall be. Such models are inspected in the clay, re-
touched and worked over until the architect is satisfied, and
when approved, the glue or gelatine is poured over them in
preparation for the making of a plaster counterpart. The making of moulds for large and complicated ornamental compositions is again part of the technic of terra-cotta making, and the
possibilities of reproduction must be known by the modeller.
Manufacturers are alert to eliminate "undercut" ornamenta-

Courtesy of Atlantic Terra Cotta Co.

Clays from which building ceramics are obtained. The mining is carefully supervised to insure uniform quality.

Courtesy of Kittanning Brick & Fire Clay Co.

Making hollow tile fireproofing. The clay is ground and mixed with water, excreted through the dies of the machine and is cut rectangularly.

tion on account of the difficulties that ensue in the mould-making, and in the pressing room where the next step occurs.

In the pressing room, the carefully prepared clay of just the right consistency is pressed into the moulds, the backs of the pieces being scooped out by the hands of the pressers. These men know just how thick the walls of the pieces must be and just the amount of ribbing that must be left to support the finished pieces for handling and shipping, and for setting into their final places in the structure they are to adorn.

After pressing into the moulds, the pieces are allowed to dry for a few days and are then carefully removed from the moulds. At this stage, the pieces will stand careful handling and they are removed to where the "slip" is applied. Now the slip is to terra-cotta what the enamel is to the china plate. Beyond its beauty of form, all of the beauty of terra-cotta lies in its slip; also its durability, assuming the piece has been properly designed and made. The slip is, of course, only applied to the exposed surfaces; that part which beds into the wall being left as it comes from the moulds.

The slip is generally sprayed on in modern terra-cotta plants, although where intricate polychrome work is done it must be applied with a brush much as a china painter paints china; but, of course, on a scale commensurate with the large work we are discussing. It is in the preparation and application of the slip that the greatest development of terra-cotta making has been achieved in late years. The observer can get no idea of the color or texture of the slip before it is fired. Gold may look like brown paint, and colors widely different may look much the same as they are being applied. The ceramic engineers and chemists work entirely by the chemistry of their product, serenely sure of the ultimate appearance of the material they are making.

When the slip has dried sufficiently to permit of handling the pieces, they are carefully conveyed to the kilns, where they are as carefully piled, with interstices between them to allow

Terra-cotta modellers at work. Left, the noted modeller, Angelo Ricci.

A large terra-cotta arch is being assembled and fitted. From this position it will be carefully packed in freight-cars or on trucks. Shipping lists are carefully made and each piece checked and re-checked as it is loaded, and the same care is exercised at the building upon the arrival of the material.

the heat of the kiln to circulate freely around every piece, also to insure that the slipped surfaces shall not touch, for they will fuse together under terrific heat if they are not thus separated.

And now the kiln is loaded, piled to capacity with pieces ready for the firing. Manufacturers of terra-cotta try to put as much material as possible for a single job in one kiln, first because it is an assurance of uniformity of color, and also because it is easier to keep track of the work, for the marking, recording and checking is an important part of the business of terra-cotta making.

For five days or more the fires in the kilns burn night and day, maintaining a continuous heat of about twenty-two hundred degrees Fahrenheit, a heat that would keep steel cherry-red or would melt ordinary commercial glass. The heat may be produced from coal or coke, or from gas or crude oil. Whatever the fuel, the supervision of the kilns while firing is one of constant vigilance. Pyrometers inserted at proper places in the kilns warn of slight temperature changes, and when these occur, corrective measures are taken to insure the necessary uniformity and constancy of the kiln heat. Small observation holes permit of frequent visual inspection of the material as the firing goes on.

After five days have elapsed, the fires are allowed to die out, and for a couple of days the kiln cools off gradually until the finished material is sufficiently cool to permit of handling. Again an inspection; faulty or defective pieces are rejected and notation for quick replacement made. Where there is considerable repetition of a single diagram, a few extra pieces have been made, so that, unless the occasional defective piece is some special of which there are no duplicates, the "overs" generally take care of the loss and there is no interruption at the building.

We sometimes see the façade of a building under construction scarred by a gaping hole, the obvious location of a piece

Courtesy of Federal Terra Cotta Co.

Finished terra-cotta being set up and fitted preparatory to shipment.

Courtesy of Atlantic Terra Cotta Co.

Setting the terra-cotta column capital at the 31st story level of the New York Central Building, New York.

Courtesy of Federal Terra Cotta Co.

Blowing the slip before burning. This terra-cotta is pressed and dried but not burned. The operator is coating

of terra-cotta. This means that, through some defect or accident at the kiln or in transit, the missing piece was not available when the masons were at that point in the work. The builder knows in advance of this shortage from his expediters and receiving clerks, and simply leaves the place for it to be filled in at a later date—generally when the masons are cleaning down the building shortly before final completion. Such occasional omissions do no harm. The piece when set will fit snugly into place, and the firm anchoring of the adjacent pieces insures a solidity to the whole work, of which the belated piece, when set, becomes an integral part.

From the kilns the pieces go to the fitters, who place them carefully together about as they will occur in the wall. Any little trimming or fitting that is to be done receives attention here. The material is now ready for shipment, and its further progress is merely the commercial operation of checking, listing and billing.

The expediter from the builder's organization has known of the material's every step from approval of drawings to final checking, and his report to the main office is likely to carry between its lines a sigh of relief; for there is no other material entering into the construction of a skyscraper that passes through so many specialized processes arising out of the particular need of the structure for which it is made.

It takes from six to ten weeks after drawings are completed and models approved to obtain terra-cotta. Therefore, it must be apparent that one of the early decisions must be the terra-cotta design, the exact extent to which it will be used, its color, form and special modelling. Builders endeavor to obtain these decisions as soon as the structural steel questions are out of the way. The proper logistical handling of this material is of special importance, for it is bulky and easily damaged, particularly where fine ornament is involved, and therefore, terra-cotta should be subjected to the minimum of handling. A carefully prepared schedule will permit of handling it from kilns to cars, thence to the job ready to lay it out and sort it on the floors immediately behind the places where it will be used in the wall.

Terra-cotta is being used increasingly for beautiful interiors, and when so used, it is scheduled with the interior finishes, such as marble and bronze. In this case it is not so vital to give it a place in the early requirements, but the inevitable six to ten weeks should never be lost sight of.

The material is set by bricklayers, not stonemasons as one might suppose. The cutting and fitting also must be done by bricklayers.

With this brief outline of terra-cotta we will leave the exterior of the building for, while there are many other important operations in completing the enclosure, they cannot claim the special interest that attaches to the more specialized operations which have become identified with our skyscraper construction. Roofing and sheet-metal work are prosy, work-a-day operations, unless we have a large sheet-metal cornice to hang. The roofing itself is done with care, and the science has been so far developed that, given a good specification with sufficient appropriation to secure a thoroughly first-class job, the work causes the builder little concern. The best flat roofs we see are constructed of five-ply tar and felt on a cement surface, as well laid as a smooth sidewalk. On top of this water-proof membrane is generally laid the tile surface we see—pleasing and durable, for roofing tile has come to be a standard article, vitreous and everlasting if properly treated. In and through the parapet walls, which are usually more than three feet high above the general roof level, copper flashings have been built. The expert roofer knows how to seal the membrane to the flashings and how to make everything securely water-proof, if he is but given his way. Where surfaces are large—more than fifty to seventy-five feet square, let us say—expansion joints are introduced by constructing a hollow sheet copper rib or seam across the area, and this is sealed to the membrane by a hot mopping of tar, just as the flashings are. There are many ways to accomplish this allowance for expansion, and the one here given is by way of illustration.

Where sloping roofs occur, copper may be used, and if it is, great care must be exercised at all hips and valleys and at the edges, for copper has a way of weaving and tugging at its moorings under the action of heat and cold. Sheet lead roofs are of ancient origin, and were used by the cathedral builders of old. The roof of the Cathedral at Cologne was ripped off at the direction of Napoleon and cast into bullets for his army, but long after the passing of Napoleon, the roof was restored in lead, and even to this day it seems to be the best roofing available for the great cathedrals throughout Europe. In our American cities, lead seems to yield to the action of the elements. Perhaps this is due to the peculiar acidity of our smoky city atmosphere. The subject is still controversial between the advocates of various roofing materials.

Many architects are specifying lead-coated copper for sloping roofs, on the theory that, by this method, the pleasing appearance of lead is obtained while at the same time the greater durability of copper insures a more lasting roof.

Tile roofing is also ancient and also very good, and its method of application is about the same as it has been ever since such roofs were first used.

Sloping roofs are being used increasingly by architects of these tall buildings, due in part to the setback laws and in part to an increased appreciation on the part of owners of the requirement to house-in water tanks and unsightly excrescences on the roof tops. The development has added a new note of interest to our large cities, a commendable variation in our national contribution to design—the skyscraper.

Moulders pouring the molten metal for a piece of beautiful interior bronze work.

Part of the great bronze lantern that surmounts the New York Life Insurance Company Building, New York.

Model for a bronze door, from which the moulds are made.

CHAPTER XVIII

INTERIOR STRUCTURE

WE have seen that the enclosure of the building is a turning point of great importance to the builder and a major goal toward which his efforts have been chiefly directed, but after all, it is only a milestone, and long before the steel has reached the top, much of the material that goes to make up the interior has been ordered and is well under way. Immediately after signing the contract for the work, the builder turns his attention to the all-important business of subletting those parts that, through trade practice and the nature of the industry, must of necessity be sublet. Certain items, such as elevators, cut stone, terra-cotta, marble work, millwork or steel interior trim, must be sublet, for no builder has within his organization the facilities to do these lines of work. Structural steel must be purchased from the fabricators for the obvious reason that the fabricators' claim to economy of operation arises from an annual tonnage far greater than any one builder could ever hope to use. The builder may elect to set the steel, but excepting in rare cases, the principal justification of this is the desirability of controlling the progress, for, as we have seen, the time schedule revolves about the delivery and completion of the steel structure. Likewise, plumbing, heating and electrical work are almost invariably sublet, for a builder would have to command a prodigious volume in order to absorb the overhead and shop capacity that govern the economical installation of these lines of work.

Trade custom is a pretty safe guide. Some lines proclaim

themselves as necessarily to be sublet, as we have seen. Others have been found by long practice to be most economically performed by sub-contractors, even though the builder may have the necessary qualifications to do the work; and some lines of work, such as cement floors, concrete floor arches and, occasionally, carpentry, are even yet in the twilight zone where it may be advisable or expedient to sublet them on one job, while, on another, the builder would be well advised to perform the same work direct. It is in the clear judging and proper disposition of these matters that the capable builder excels.

The floor arches are a necessary accompaniment to the structural steel, first because building laws almost universally require that they follow within two or three stories of the floor on which the derrick works, but beyond this because they are useful places of storage for materials that swiftly follow the steel. Builders can remember when this was not so, and as late as 1900, steel skeletons of some very large buildings were erected complete before even the first floor arches were laid. The accidents to the steel erectors were, of course, proportionately higher, and the seemingly unnecessary casualties of such operations brought them under the proper scrutiny of city authorities, the result being the laws that, as has been said, are now very generally in force.

But to return to the floor arches The spectator thinks of an arch as that graceful curving part of the structure generally found over doors and windows, and it is something of a tax on the imagination to contemplate the flat slab structures between the floor beams of a skyscraper as "arches." We have seen that the earliest endeavors toward non-inflammable structures were by means of brick arches actually put up in accordance with the orthodox notion of such construction when it was in vogue, and we have also seen that Kreischer, through his invention of the flat arch of hollow tile, greatly improved the original masonry floor arch. Every layman, of course, knows that the flat arch holds a normal and legitimate place in architecture, and it takes no stretch of the imagination to perceive one of the principles of Kreischer's patent that had been established by its predecessor in the wall, the Jack arch. A Jack arch gets its name from the resemblance to the crown on the jack in a deck of playing cards. In the orthodox deck, that whimsical head-piece looks like the construction of certain window heads that are sometimes seen in either stone or brick, although the similarity is most apparent in a brick Jack arch.

The hollow tile principle having been accepted, it came to be carried out in two ways; first, where the flues in the tiles ran longitudinally or parallel to the floor beams between which the arches were built. This is called a "side-construction" floor arch, and is easily recognized by the seeming lop-sidedness of all the pieces excepting the keys, which hold the symmetrical form characteristic of an arch key wherever found. The second method, known as "end construction," derives its name from the fact that the flues in the tile, instead of running parallel to the supporting floor beams, run at right angles thereto, the ends of the webs coming in contact thus forming a series of flues transversely through the floor arch. This latter method has the manufacturing advantage of being cut from a single rectangular column of clay as it is excreted from the machine, the wire cutting being made at various angles to conform with the radial lines of the arch of which the various shaped pieces are to be a part. Standardization, both in arch construction and in spacing the floor beams, has simplified this question of the radial cuts, or slopes; for in either construction, a beam spacing of from five to six feet can be met by the same radial formation, because any discrepancies from true parallel of the radial surfaces will be made up by the mortar in which the arches are laid. The tile that fills into the beam at the starting of the arch is called the skew-back, and the intermediate blocks between the skew-back and the key are sometimes called voussoirs. The key is what its name has always meant when arches are discussed. The argument still rages as to whether side or end is the better form of construction. The fact is that they are about equally good and both are very much stronger than would be required for the ordinary loading.

Any consideration of floor arches must include the ingenious development of Guastavino, that interesting artist engineer who, a half-century ago, started experimenting with ceramic tiles, of which he was a past-master in the making, with the view of using them structurally. The advent of skeleton construction gave the impetus needed for his work, for, although the laminated arch construction which he carried to such complete and artistic solution had been used prior to his time, the development was his, and already the Guastavino arch has almost generic significance in describing a type now so generally used, where the combination of a beautiful ceramic, domed or vaulted under-surface, combined with light weight and thinness of construction and great strength are desired. The under side of any arch resisting gravitation is called the soffit. Guastavino soffits are known the world over and may be recognized by their pleasing use of the simple forms of which they are made. Generally about four by eight inches, plainly rectangular in form, they are laid sometimes in running bond, sometimes in herring-bone pattern, sometimes interspersed with beautifully colored ceramic borders or patterns; always fitting and beautiful in the hands of a competent architect. The Guastavino arch gets its strength from the camber of its form. The individual tiles are generally less than an inch thick, and they conform to the dome or vault ceiling which they are to form by the slight changes in plane of the individual pieces as they form the concave surface we see. Cement mortar joints between the tiles permit of this, and after the work has been laid and the mortar set, the joints are pointed up, giving a mesh-like pattern to the surface.

Where a barrel vault or a groined ceiling is involved, a centre or form must first be laid, in principle, the centre that any arch demands until its keystone is set. The work is started by laying a lamination of these tiles with their predetermined ceramic color scheme face down on the centre. One layer having been laid, the masons go back over the upper surface of it with another layer, sometimes called "filler" tile, just as hard-burned, but of course, omitting the costly enamel surface. This second layer is so laid that the joints "stagger" with the first layer, and being bedded on the first in a layer of Portland cement mortar, the construction takes on the form of a homogeneous whole. On this second layer a third is laid. By this time a shell of exceeding strength has been developed, the number of laminations, of course, determining the strength. Ordinarily three layers or laminations are all that are necessary. Where very wide spans are involved, and the work can be done in spans up to fifty or even seventy-five feet, the laminations are increased, occasionally up to seven or eight layers in the great domes or vaults we sometimes see. Where a spherical dome is to be built, it is possible to do the work with only the rudiment of a centre, just enough to get a few circles of tile started out from the edge. With this start, the skilled workmen will lay ring after ring around the dome until it is finally closed and keyed with a last piece or an appropriate centrepiece, from which a great chandelier may be hung. Such work must be done with great care and with due regard for the form. This is accomplished by having ribs or templates cut in the form of a segment of the dome, every course of tile accurately tested by this template.

While Guastavino arches may be used most frequently in decoration, primarily they are structural, and the ingenuity of their construction, so different from all other forms of arch,

A gyratory crusher used to break up the limestone or cement rock used in making Portland cement. Lumps of rock three feet in diameter are handled by this machine.

A rotary cement kiln, the largest type of revolving machinery known to industry. Powdered coal is blown into the kiln, igniting a 40-foot flame and making a temperature of 2700 degrees Fahrenheit. Using colored glasses, the man at the front watches the process of calcining of the raw materials and regulates the

gives them a unique place in the science of modern building. It was the development of the ceramic art and of Portland cement that made this ingenious construction possible. Guastavino arches have been used extensively in our finest public buildings and in the newer railway terminals. The North Western in Chicago and the Grand Central in New York are decorated with beautiful examples of this work.

The invention of reinforced concrete ushered in the concrete arch which, in some localities, has almost replaced the hollow tile arch. It is nothing more than a flat slab of concrete —a stone cast in place—with reinforcing rods or mesh imbedded in it. The concrete is made to cover the sides and soffits of the beams that support it for fire protection, but the strength is obtained by the combination of the high tensile strength of the reinforcing and the compression strength of the concrete. Cinder concrete is largely used in cities where quantities of cinders are available, both because of its lightness and its relatively lower cost as compared with crushed stone, slag or any other filling material. Stone concrete is heavier and the stone ingredient costs more, but the weight is the main deterrent from its use as it runs up the tonnage of structural steel by reason of increasing the dead load. In some sections of the country, blast furnace slag is available and relatively inexpensive. It is of light weight and altogether an almost ideal material.

This question of the weight of the floor arch is ever on the mind of the structural engineer, and inventors continue to attempt all sorts of ingenuities to make the floor construction lighter and thus reduce the amount of steel to be used. One system goes so far as to use plaster of paris and shavings for a filler and compression member, the reinforcing, however, being somewhat increased. But whatever the method, the object is always the same: to get a strong, light floor construction.

Sound-proofing must have consideration but is secondary in the ordinary commercial work, and indeed, the question is a difficult one. People with sensitive ears may demand elaborate precautions and still not be satisfied; others will not notice sounds from adjoining quarters. Deadening materials of various sorts have been used in floor filling and in hung ceilings, and yet the matter of the amount of sound transmitted and its relative annoyance remain in the realm of personal opinion. Suffice it to say that almost any of the standard floor arch constructions transmit to a certain extent the sounds from above, and when these cannot be tolerated there is no effective half-way stopping place between taking matters as they stand and elaborate constructions designed effectively to stop sound transmission.

Portland cement, that essential item in skyscraper construction, takes its place with structural steel in the stimulus furnished to intensive, metropolitan, building development. We have seen how indispensable it is in foundation construction and how valuable its service in almost every phase of building work. The material was invented in England' by an English mason, Joseph Aspdin, in 1824. He called it Portland because concrete made with it resembled a building stone extensively used in England which is obtained from quarries on the Isle of Portland, off the English coast Many of the great structures of England are built of this stone, among them Westminster Abbey. Aspdin could have had no realization of the extent to which his discovery would be carried, any more than Bessemer could have visualized the skyscraper when he first invented his process for refining steel out of pig iron. But the development has come nevertheless, and to-day Portland cement is an indispensable factor, not only of construction, but of modern life itself. No less than one hundred and seventy million barrels are manufactured annually in this country in one hundred and fifty plants located in thirty different states.

The process starts with a cementaceous limestone quarry where rock of the proper quality is mined or quarried and sent to the crusher. It is a prodigious operation, for the rock comes there in great chunks larger than a man's body, and is fed into the maw of a huge rotary crusher that grinds on incessantly and insatiably. Every few seconds a huge chunk of rock comes to grips with the inexorably crunching maw. The outcome is always the same; the rock is caught, broken, and drawn down into the crater, the huge machine trundling on unperturbed. The crushed and defeated rock, now in lumps the size of a man's two fists, pours out of the crusher onto a belt of buckets that raise it and dump it into another crusher. This time it is reduced to the size of the two-inch crushed stone of commerce, and at this stage, it would make an excellent aggregate for concrete, but the material is destined to play a far more important part in that indispensable product.

Crushed to this smaller size, it is moved on by another continuous bucket belt, now to find itself spilled into a belt conveyor running over a huge bin house where it takes temporary repose, while chemists dip in and sample its composition, that the estimate of its purity and fitness, which they had made before the quarry was put in operation, may be confirmed. The rock is of slightly variable composition as it comes to this great bin house, and the tests are frequent so that any deficiencies or variation may be made up in the next step, for there the rock is to meet its first blending operation.

Portland cement is not just rock treated in a certain way, but a scientific admixture of accurately proportioned ingredients, of which limestone, while the principal one, must have with it a proportion of clay, either shale or blast furnace slag, lime, silica and alumina. So in this first great bin house, the chemists determine what proportions of these ingredients must be mixed to insure the even quality of the cement.

From the bottom of these huge bins, which hold perhaps

two hundred tons of crushed rock at all times, the sampled rock is dropped on a conveyor belt. On that same belt material is poured from other storage bins, shale or blast furnace slag and clay, all in correct proportions. The mixture, always moving onward, is again bucketed up to another bin, from which it is chuted to a great horizontal revolving drum. This drum at once thoroughly mixes and further grinds the aggregate, this time by the use of tons of steel balls about the size of large marbles, and these do their work by tumbling on the mixing aggregate as the drum is revolved. The aggregate, now thoroughly mixed, again travels by bucket to the raw-mix storage bin, for now in its powdered form it is ready for burning.

On it goes through feed bins, which simply act as a reservoir, into the mighty rotary kilns in which the burning takes place. These great rotating kilns, through which an automobile could drive with ease and which are as long as three Pullman cars placed end to end, are the largest pieces of rotating machinery known to industry. The mixed raw aggregate, as it pours into the rotating kiln, meets a hot column of air rushing over the tumbling mass, and as this giant tube is pitched slightly downward, the material tends to move onward toward the oncoming flame from the powdered coal, or sometimes crude oil, that is rushing on to meet it. At first only heated by the gases of combustion, the raw mixture soon encounters the terrific flame of the combusting fuel. Driven on, the action of fusion takes place, for the heat is now run up to about three thousand degrees Fahrenheit, and only the heavy fire clay lining saves the steel sheet of the kiln from melting.

And now we have a new product—cement clinker, a glasslike material, thoroughly burned and chemically fused. It tumbles out of the lower end of the rotating kiln red hot, about the size of large pop corn, into another great cylinder, the function of which is slowly to cool the clinker. This cylinder is likewise pitched, and the cooled material is gradually poured onto a belt, to be conveyed to a large clinker storage space. At this stage, the burned material is inert. Only after it is finely ground does the clinker become cement, but the process hurries it on to the last grinding, where again great drums or cylinders, with tons of balls in them, rotate the clinker in the final grinding. It is here that the material receives that amazing fineness that makes our American cement the marvel of the world. Here, in the grinding, a small quantity of gypsum is added, just the proper amount to insure the correct setting time of the finished product, for all along the line, chemists have been sampling, and observations so made fix the amount of gypsum to be added. Ground in this fashion, the cement becomes finer than the finest flour. About eighty per cent of it will pass through a sieve with forty thousand holes to the square inch—water will not run through such a sieve.

Packed in cloth bags weighing about ninety-four pounds each, four bags to the barrel, the cement is stored in large warehouses awaiting shipment. Here the last tests are made by the vigilant chemists. Samples are taken from identified lots of the cement for many kinds of tests, the most common ones being for tensile and crushing strength. At this point also, both government and private inspectors are to be found sampling and making up little briquettes from the samples. These are tested after one day, seven days and twenty-eight days. The last of these having been satisfactorily met, the material that has been in the warehouse is now of a known quality, and its behavior at the building site is largely the matter of careful mixing and using. The manufacturer has done his part, and the responsibility for inferior results, if they come, must be due to some cause beyond the control of the maker.

All of the foregoing has been given in an endeavor to indicate to the reader the complexity of the process and the vigilance that attends cement making. The mixing and blending

A lime quarry. Lime will always have its important uses in building construction. Some extent of the demand is indicated by the enormous scale on which the material is still quarried.

Obtaining gypsum rock that is calcined and ground into modern wall-plaster.

is all a part of the technic of the cement maker's art; different
manufacturers have their own special details of process upon
which they may lay claim to individual superiorities, but the
main appeal of every manufacturer is uniformity and reliability
of product, absolute vigilance over the material in process, and
surety of delivery service, for a great construction operation
once under way cannot be interrupted by uncertainty of its cement supply or any threat of lack of uniformity of this essential
of modern construction.

Cement makers tell us that the calcination of the rock,
meaning the burning process, eliminates from the structure of
the natural material the water of crystallization, which is another way of saying that the heat de-crystallizes the limestone.
But the heat also compels a recombining of a number of separated elements contributed by the marl, slag, alumina and
other ingredients in the terrific holocaust that is created in the
great revolving kiln. In fact, a wholly new material is created,
and when ground to irreducible fineness, after the burning, it
is the Portland cement we know.

When cement is used in concrete or in mortar, or in fact
wherever it is mixed with a proper proportion of water, the
crystallizing tendency of the material, inert when dry, immediately commences to assert itself. The water temporarily
dissolves the finely ground particles, and the molecular action
of crystallization at once sets in. The crystals will attach themselves to each other and, even in the presence of the water
which at first caused the dissolution, the crystals commence to
take form and harden. Sand is mixed with the cement, first
because it is an inexpensive as well as an excellent filler, and
further because the hard facets and sharp angles of the particles of sand form ideal surfaces to which the crystals of cement will attach themselves. In like manner, crushed stone is
used in concrete to save both cement and sand, at the same
time furnishing a filler that is harder than the cement itself

or the cement and sand combined. Half or three-quarter inch stone is used in concrete that is to be laid in thin layers of complicated forms, or where much reinforcing is to be used to insure the aggregate getting into all interstices and completely surrounding the reinforcing. In heavy foundations or where concrete is massed, two inch and even larger stone is used. In very heavy foundations such as bridge piers or dams, great pieces of rock, even up to derrick sizes, are sometimes used. When this is done, a heavy bed of two-inch stone concrete, perhaps three feet thick, is first deposited, then the large stones, first thoroughly washed and with clean split faces, are dropped into the plastic mass, great care being taken that they are fully bedded and in the centre of the mass. Again the succeeding layer of concrete fully covering any projecting points of the large stones is deposited, and again the large stones. Such concrete is sometimes called "Cyclopean concrete."

The microscopic crystallization process of the cement is not unlike the crystallization of sugar. Whoever has seen a bucket of rock candy in the old-fashioned drug store window has seen something like the sight that greets the eye of the observer as he gazes through a microscope at a cement crystalline formation.

We hear of brick mortar being specified one to two or one to three "tempered with slacked or hydrated lime," which, of course, means one part of cement to, say, three parts of sand. The "tempering" is the admixture of a material that makes it work "smooth" under the mason's trowel, but when properly used, it has the effect of adding a degree of moisture proofing to the hardened mortar, for as the cement sets in the presence of the lime in suspension, the lime has a tendency to fill the interstices between the crystals, and moreover, is itself a binder between them. Little can be claimed for any added strength. In fact, tests show that lime measurably weakens the cement, but the material is excessively strong and can afford the slight reduction in strength that the use of lime entails because of the great advantage that arises from the smoother working of the mortar. In both mortar and concrete there is a proper proportion of cement and the other ingredients, yet there is a deep-rooted feeling in the minds of some architects and engineers that the "richer" the mixture the stronger it is. Such is not the case. The cement is the least resistant of the ingredients of concrete, and under fracture tests, it is the cement that first gives way. Neat cement will not resist the same crushing that either sand and cement or sand-stone and cement will. People gaze with smiling admiration at the efforts of workmen to break apart some old concrete structure. It is, of course, a tough job, but in reality it is the ever-interesting phenomenon of cement that attracts them. Concrete will yield to the drill and sledge hammer much more readily than will the original rock from which the crushed stone came.

Concrete, while almost ideal for many forms of construction and particularly for foundations, must not be considered the everlasting and infallible material for all possible uses. One has but to observe the concrete retaining walls along railways and he will occasionally see concrete, new and old, rapidly deteriorating. It is common to see the outside facing scale off, and sometimes the body of the wall itself is seen to be almost chalky in places. This is not because the concrete was poorly made originally, but because the chemicals, particularly the salts and alkalines of the ground, carried in by capillarity or by seepage, actually break down or sometimes dissolve the crystalline structure of the hardened cement. It is an insidious and baffling thing, and as yet full means of combating it have not been developed. The railroads, on their more recent work, have resorted to water-proofing with greatest care the backs of all their retaining walls, thus forestalling the possibility of moisture entering the concrete at all. This, however, does not seem to be the full solution, for the joints between the sections of the work and even the lines between batches, particularly if a few hours have here elapsed, seem to contain the initial elements of a disintegration that is very hard to combat.

Concrete underground, where saturation takes place and then comes to rest, does not show any of these signs of disintegration. Neither does concrete under roof or not exposed to continuous or intermittent moisture show this weakness. It will be seen from the foregoing that concrete is not just a mixture which, once made, will be everlasting regardless of its use or abuse, and its environment. It is a subject of deep study by specialists, and care and understanding must attend its use. The American Concrete Institute, collaborating with the Portland Cement Association, carry on an intensive study of the problems involved. They have accomplished much and have yet much to accomplish.

Of recent years we have been afforded a quick setting cement which has many special uses. On rush jobs the forms of the concrete arches can be struck within twenty-four to thirty-six hours after the concrete is placed. Ordinarily the forms would have to remain in place about seven days. But building construction is not the only beneficiary. Road builders in busy city streets can lay a road-bed to-day and use it to-morrow. One could think of a dozen uses to which this quick-setting material could be put. During the World War the material, then new, was used to good effect in making hastily constructed gun emplacements, redouts, and machine-gun "pill boxes."

CHAPTER XIX

WE COMMENCE TO FINISH

I⊤ is not necessary to recount all of the operations of completion in a building to give a picture of the general swift progress of the work, once the building is enclosed. We have seen that the pre-arranged plan of action requires that immediately on the heels of one trade another follows, and that the builder, through his organized forethought, has arranged or sub-contracted for everything that goes to make up the completed structure. The watchword is to keep driving, everlastingly driving, all with due regard to the quality of the work and the sequence—inspecting, planning, conferring, working out details of method, adjusting minor inconsistencies. For the plans, however perfect, cannot possibly depict every contingency, and the broad generalizations of the specifications properly require co-operation and co-ordination; these are the functions of the building superintendent, always alert and always looking ahead.

Interior partitions are of masonry and laid by bricklayers. The superintendent is apt to have directed that the best men from the outside walls be kept and thrown in on partition work. Frequently inclement weather forces a slowing down of the exterior walls, and the competent superintendent has seen to it that some of the floors are stocked with partition tile against this rainy day. The builder knows this also, and has been pushing the architect for any special layouts to be built in floors where the exterior walls are finished. But further description here becomes too complex. The interior requires that interlocking action of steam-fitters, plumbers, electricians, sheet metal workers, elevator constructors and ornamental iron workers, to say nothing of as many more essential trades, with the progress demanding that no nook or corner of the job be idle while work is there to be done.

In the earlier days of skyscraper construction, partitions were almost universally of hollow tile, sometimes three inches, sometimes four inches thick. Six and even eight-inch partition blocks are made, but these are for special uses, such as the legal requirement in some cities of thicker partitions around elevator shafts and stairways. Ceiling heights of fifteen feet and over require at least six-inch partitions if good practice is to be followed, although columns of almost any height may be covered with either three or four-inch tile, unless they are concreted in and no tile is necessary.

Of late years, gypsum blocks have successfully competed with hollow tile, and in some cities are so considerably cheaper as to dominate the market. When well made, these blocks are very satisfactory. They are lighter than tile, come in larger sizes, can be cut with a saw, are straight and uniform in size and thickness; nails, easily driven into them, take hold with a tenacity almost equalling that of nails driven into wood. In spite of these many advantages, architects of quality structures are apt to stick to tile specifications for interior partitions. They are somehow regarded as more substantial, have some advantage perhaps in the better adhesion of the plaster, and never shrink, thus avoiding the unsightly crack along the ceiling that sometimes is seen when inferior or "green" gypsum blocks are used.

Metal lath partitions also are used and, in some cases, are indispensable where a very thin partition is required.

In partition work, the architect allows for a very exact finished thickness, and any deviation will cause trouble at door jambs and around windows, whether the trim be of wood or metal. In either case, the trim is prepared in advance from full size details that guarantee to the trim manufacturer that an exact partition thickness, plaster to plaster, will be found when the trim reaches the job ready for erection.

Whether the partition blocks be of tile or gypsum, the next operation is putting on grounds to furnish a guide and set the plane of the plaster. Carpenters do this work, for the grounds are generally of pine or spruce about a half inch thick and a couple of inches wide, one generally found at a line about six inches or so above the floor, as the grounds also serve as nailing strips for the finished base that is to follow. Another ground is put on at the line of the picture mould, again serving the dual purpose of establishing the plane of the plaster and furnishing nailing for the picture mould. The full sized details have furnished the information as to just how and where these grounds are to be placed; the woodwork manufacturer finds them just where the drawing showed them, when he arrives with his material some time after the plastering is finished.

If steel trim is to be used—and we are using steel trim and doors more and more—the whole doorway is set before the partition work starts, and is solidly built in by the masons, for the steel needs no unusual protection and it does no damage to have the mortar of both tile-laying and plastering smeared all over the steel. All that is necessary is to have it cleaned with a wire brush after the mortar work has passed, preparatory to final painting.

Recently, metal grounds have found considerable favor, especially where metal trim is afterward to be installed. It is an ingenious construction and has its advantages, but presents a problem of fastening the succeeding trim. The problem of base has best been met by installation of cement or other form of plastic base material, in which case the metal ground is ideal and the wood ground guide objectionable.

The completion of the plastering is the goal of the builder

in his consideration of finishing the interior, just as the completion of the steel was his goal in the exterior, as we have seen. It is messy business and everything stands aside when the plasterer, organized for swift progress, sweeps through the building. Floor after floor is given over to him as he progresses, generally from the top downward, taking his clutter of scaffolding and the mess of his waste with him. Floor after floor must have been made ready; everything—pipes in, grounds completed, corner beads set, electric light outlets accurately placed and exactly at the plane of the plaster as established by the accurately set grounds; every one endeavors to finish his roughing work and flee the clutter and mess that this oncoming crew inevitably brings with it.

Consideration of partition work and the succeeding plastering thereof holds a very important place in the calculations of the builder. The anxieties are not the same that attended the foundation work, for there safety and the everlasting standing qualities of the structure were involved. Now the builder is concerned in excellence of finish, for, as we have seen, he may have done the foundations and structure well, but if the interior does not answer to the critical requirements of neat and careful workmanship and excellence of finish, the whole operation is apt to be adjudged inferior. So straight true plaster work must be done, and this depends in a large part on grounds and corner beads, but not altogether, for the material of plastering must answer to the requirements of speedy application, quick setting and drying, and the reasonably rapid succession of the coats of plaster.

Plastering, from time immemorial, has been done with lime mortar. All recorded history of building shows that early in the dawn of civilization, lime plasters and mortars were used. Even in the early days of the skyscraper, one of the problems to be met was the setting apart of a large area in the basement where, soon after the steel was set, quantities of lump lime would be brought to the building to be slaked. We all remember the familiar sight of the mortar box with its lumps of slaking lime, and the consequent heat that seemed to cause the water to boil and bubble as the calcium oxide, released by the chemical action of the water on the lime lump, gave off its gases and, in fact, generated great heat in the process. The lime putty so formed, but now inert, was "fat" or lean, according to its origin, and the content of impurities. There were high calcium limes and dolomitic limes and marble limes, which were nothing but variations on the two basic branches of the lime family first named, but every plastering foreman had his prejudices and his theories. Fat, smooth-working, thoroughly slaked and inert mortar was his objective, and how to get it was the debated question.

The space that the older method required, together with the uncertainties and carelessness of job labor and the demand for a surer and more quickly usable material, resulted in the introduction and perfection of gypsum plaster, and to-day that material is used almost exclusively in the brown coat, with lime putty still the ideal material for the finish.

Gypsum plaster is made somewhat in the manner of Portland cement, but the process is vastly simpler. Gypsum rock is mined, and after being crushed to convenient size, is dried and calcined by what amounts to a simple operation; that is, the crushed material is fed into one end of a kiln and, by means of rotation or of rotating paddles, is advanced toward the source of heat, generally coke, gas or crude oil. As soon as it enters the heat, calcination—which, as we have seen in the case of Portland cement, is simply breaking down the crystalline structure of the rock—commences. As it reaches the greatest heat, the calcination is completed and the hot chalk-like substance so formed falls into cooling bins, and is belted on from there to grinding machines, either of steel balls, as in the case of Portland cement, or by other process, that reduces it to a suitable fineness. This material is reasonably soluble, and while in the presence of water it tends to reform, its hardening, unlike cement, is accomplished largely by its drying. Gypsum plaster will dissolve under the continued action of water, and in this respect has nothing in common with true Portland cement.

The development of gypsum plaster was a boon to the skyscraper builder, and immediately after its introduction, it took its important place in the industry. Gypsum plaster is delivered either "neat" or mixed with sand. The latter is frequently specified because of the accuracy of the sand mix that the mills provide by automatic machinery. Neat plaster delivered to the job has the merit of cheapness, for the plasterer can buy sand about as cheaply as the gypsum manufacturer can, and by getting it direct, he saves haulage and handling, for which the gypsum manufacturer must charge in the price for the mixed product.

And now the gypsum plaster is delivered, and for a week or more the job is all agog with the receiving and hoisting of the bagged material—so many tons to each floor, for it will be seen that here is storage space of a sort. Temporary water lines that were run up through the building as the outside walls were built now serve the plasterer. Mortar boxes are distributed about at convenient points on each floor, and with all this preparation in good order, on the appointed day the plasterers commence to arrive. Their swift course through the upper reaches might ineptly be compared to the sweep of a swarm of devouring locusts; but they do not devour, they construct. In a few days, the whole appearance of the floors commences to change. First ceilings and side walls down to scaffold-high are done. Then the well-trained crew of scaffold builders, plasterers' laborers, appear and whisk away the scaffolds, while the plasterers themselves have moved on to another floor. The scaffolds out, back come the plasterers, seemingly in perfect team work, and finish browning the walls from scaffold-high to the floor.

A period of quiescence sets in. Plumbers, electricians, fitters and marble setters commence to appear, prodding about in the wake of this crew that has passed through the building, to take up the tasks left undone until the plasterers had finished their browning. Plasterers' laborers follow on the floors after the plasterers have left and commence cleaning up, and it is indeed a prodigious amount of cleaning they do. Tons on tons of now dead and inert material, the droppings from the main operation, are shovelled up and dumped into the chutes provided to keep the operation free of rubbish at all times. The waste seems enormous, yet it is unavoidable. Rule of thumb builders used to say that they could figure the tons of plaster required for a job by adding one third to the tons of steel it contained. Thus, a job that had three thousand tons of steel would require four thousand tons of plaster. This method of calculating is not recommended, but it gives occasion to quote the second rule of thumb, which says that a third of the tonnage of plaster that comes in, goes out again as rubbish.

Where metal lath is used—and much of it is used where hung ceilings and a considerable amount of ornamental plastering are called for—it is necessary first to plaster the metal lath with a "scratch coat." This is the same material of gypsum and sand, but with the added ingredient of cow's or goat's hair. The use of hair in plaster is as old as the use of wood lath, and its object is very obvious. On metal lath, it is necessary to get a body established that is comparable to the masonry partitions or ceilings on which the brown coat is applied. This scratch coat is done in advance of the brown and allowed to harden for a day or two before the browning starts. In applying this coat on the lath, the workmen scratch and score the plastic surface as they are about completing it, to provide a roughened surface, hence the name, "scratch coat."

Slaking lime on the job, even for white coating, soon became a hindrance to rapid building. Again its uncertainty of
composition introduced factors of possible delay, and out of
the demand for a surer and speedier finish, hydrated lime was
developed. Nowadays, hydrated lime is delivered to the job in
sacks, hoisted to the floors, turned out into mortar boxes, and
the water is applied simply to reduce the material to plastic
form. Immediately, it is trowelled into the wall with no waiting for slaking, or the tedious hoisting of putty from a great
bin in a most needed part of the basement. Hydrated lime is
simply lump lime ground fine, with water in the exact amount
to produce complete slaking introduced. The product is scientifically prepared, dried, powdered, packed in bags, and delivered—another boon to the impatient builder.

The plasterer now turns back over the job with his clutter
of scaffolding to apply the finish coat, and this time the work
moves more rapidly and the rubbish of waste, while annoying
and, as before, requiring another clean-up, is less in volume,
as the droppings from the white coating of the walls and ceilings are of relatively small amount. Again the scaffolding is removed, this time lowered to be taken away from the job altogether; a last brief turn of the plasterers as they finish the
walls from scaffold-high to base ground, and they are gone
forever, until a few of them shall be called back to do the
patching after all other trades that precede the finalists—the
painters and decorators putting the finishing touches on the
building.

Now the job takes on a very different air. Rooms have assumed their finished appearance and the most unappreciative
tyro can tell what it is all to look like. Pleased owners may
now walk about the job without hazard to their persons or
clothing, and view the almost completed work without the
risk that attended earlier visits.

It is at this point that the plumbers' and fitters' helpers

cause greatest wonder as to their reason for existence, for, as one walks from room to room, he is apt to come upon them singly and in groups, smoking cigarettes or uselessly feigning some impromptu errand. Job discipline is most difficult from now on because of the endless number of rooms and the difficulty of observation. At this juncture in the building arises the old adage that all of the losses in building come in laying the thresholds.

Operations from now on are final operations—finishing of every sort. True, the tile setter and, to a less extent, the marble setter cause a certain amount of rubbish and waste, but their work is generally confined to toilet rooms and corridors, and, of this latter, an ever-decreasing amount. Marble wainscotings in corridors are not installed as extensively as they used to be. Substitute materials for tile and marble floors are finding increased favor, high labor costs being responsible in a large measure.

It has been inferred that the plasterer browns and whiteplasters the walls everywhere, but, of course, not where tile is to be laid or marble set. Behind tile the walls are scratchcoated with a Portland cement plaster; behind marble the partition blocks are often left bare, although some architects require a thin rough plaster coat to cover them before the marble is set. Wood grounds that have been set to establish the plaster line are removed where marble and tile are to join the plaster.

But the interest now is in plaster, and with the upper floors completed, we may take time to observe the ornamental plastering which generally occurs in the main rooms of the ground floor and in the main entrance corridor. Some of the greatest triumphs of American architecture may be found in the notable banking rooms of the large cities of this country, while in the ballrooms and assembly halls of the great hotels, in theatres and public buildings may be found the crown jewels of the magnificent structures they adorn.

It is an interesting observation that rooms get their feeling of spaciousness from their ceilings, the side wall space being of secondary importance in creating this effect, even though the room be of lofty ceiling height. For this reason we Americans have come to admire great areas unbroken by any columns whatsoever, even though the arrangement of facilities such as bank screens, or furnishings such as are found in spacious hotel lobbies, crowd us into rather limited areas of floor space. Large sums of money are sometimes spent to obtain these great columnless rooms, and it is upon the design of their ceilings that the architects lavish time and money in the attainment of the pleasing effects we see.

Such a large room is a forbidding, masonry bordered area in the building under construction, stacked with material, and cluttered with sub-contractors' lockers. It is probable that a temporary hoist or two to serve the upper part of the building penetrates the ceiling area, and perhaps a roadway used by the trucks delivering material to the building crosses the floor. As soon as the pipes serving the superstructure have been tested in this area, the builder commences preparation for the ornamental plaster that will embellish the room. If there are deep girders incident to the wide spans to be covered, either the design will take them into account, or a hung ceiling obliterating them wholly or in part will be called for. Such large rooms are generally ventilated, and the ventilation ducts are installed before the ceiling is started. But, whether flat ceiling or panelled, the wire lather is soon at work hanging the ribs of the ceiling and forming the beams or heavily projecting ornamental features. This work may be done from trestle scaffolding built up from the floor, or, if it is desired to carry on the work on the finished floor at the same time, the scaffolding is hung from the floor above from hangers running through that floor, which remain in place until the plastering is completed. At this time, the hoists running through the room are boxed off or otherwise protected. The metal lathing giving form to the ceiling is scratch coated.

In the meantime, the architect has been inspecting models of the ornament. They are full size and are made in clay for his inspection, just as the models for the terra-cotta are made.

Modelling has become almost indispensable to modern architectural design, particularly that part which includes ornamentation. The large terra-cotta manufacturers have skilled modellers who accurately interpret the drawings of the architects; also the plaster contractors, bronze works and ornamental iron shops; in fact, all of those branches of building construction which minister to ornamentation have modellers on their staffs. Besides these, there are professional modellers selected by architects, who work independently of the manufacturers of various materials, and these deliver their models in the clay after they have been approved by architects.

For plaster work the modelling is always in clay, and when the model is finally finished, a preparation of gelatine or glue and plaster of paris is poured over the clay to make a mould, in the manner heretofore described in the making of terra-cotta. In ornamental plaster work there is apt to be a great deal of repetition, and much skill is involved in making models and parts of models in such a way that from them can be made the intricate compositions of ornamentation that we see.

We have seen how a formed ceiling is made of wire lath on metal ribs after the ventilation ducts and pipe work have been put in place. Now, this lath surface, having been scratch coated, is smooth plastered with cornice moulding "run" in place, and the plasterers are ready to apply the ornament. Cornices or cornice mouldings are run after the plane of the wall and ceiling has been established by grounds to form a guide for the cornice mould. This mould is cut from sheet iron of the exact profile of the cornice. The plasterers, usually working in pairs, walk along the scaffolding, one pushing the mould and the other feeding the plaster, which is highly tempered with plaster of paris. In this way, the heavy section of the cornice is built up, true and straight, with every line clean-cut. In this work the lime putty is said to be "gauged" with plaster of paris to establish the rapidity of setting of the plastic material. It is the principle of the carpenter's moulding plane adapted to plastic material. At the corners and intersections the mould is laid aside and the intricate mitering is worked out with hand tools. The ornament is generally cast in plaster of paris and is "stuck" in place with the same material, true to line and straight, in fulfillment of the architect's design.

The laying out of one of these great ceilings is sometimes a most exacting and intricate bit of work. The rooms we see are not always exactly rectangular, as they may appear to be, and the minutest variations must be taken into account. The great ellipses and curves we look up to admire, and the interesting and intricate geometric patterns reflect the painstaking work of master craftsmen. Furthermore, the lines of the plaster ceiling must coincide exactly with the lines of the ornamentation of the walls and sometimes of the floors. In such large rooms, floor, wall and ceiling must be considered as a unit, and the slightest deviation in form will throw the delicate mouldings out of line and cause distortion that the eye is quick to perceive. It is here that we require the most skilful workmen, and those who think of plaster merely in terms of applying the plain surfaces of the straightaway room must realize that these same artisans, to be all-around craftsmen, must understand the application of ornamentation as well. Indeed the best of them must have a fair understanding of descriptive geometry in order to work out the problems involved in a great, complicated ceiling.

We have said that these beautiful interiors are the crowning achievement of the skilled architect when he has an opportunity to give sway to his talents. The architecture of an interior may in no way reflect that of the exterior. Different rooms in the same building may be of totally different architectural styles, yet the skill and understanding must be there, and the better the architect, the better the interior is sure to be.

In its lovely whiteness, before any decoration is applied, a plaster ceiling is one of the most charming and gratifying things to be seen around a building. It postulates everything that is beautiful, and seeing it fresh and clean, one almost regrets that it is to be decorated; yet decoration of some sort is always necessary. The fact is that the plaster ceiling, white and beautiful as it is, is measurably porous and dirt that lodges upon it is not easy to remove. Therefore, it should be painted in some fashion, and the architect, realizing this, and also realizing that, in its beautiful plainness, it does not fulfill his conception of the design, always has in mind that it will carry a certain scheme of decoration. The decorative scheme, however, need not necessarily be so minutely studied in advance, because again we have in the decorator of these ceilings an artist himself. Some architects compose the artistic coloration of ceilings and walls and illustrate them by block drawings, but even these are only regarded as generalities. The decorator, like the modeller, is chosen for his artistic skill in interpreting the architect's design, and is apt to be given wide latitude of interpretation.

We see these beautiful ceilings sometimes appearing to be of heavy oaken beams, sometimes of marble, but more frequently seemingly of Caen stone or some other form of stone work; yet in nearly every case they are the same composition —plaster skilfully prepared by the decorator. Such imitation is entirely legitimate and indeed desirable. This is one of the few cases where imitation wood, for example, is better than the material itself would be. Likewise, beautiful forms in imitation of stone can be produced that are entirely legitimate, yet the nature of the skeleton structure itself is such that actually to produce these same forms in stone would require a prohibitive complexity of stone-cutting and setting.

It has been noted that plaster was one of the early essentials of building and has come down to us from all time. It is a perfect medium and seems completely to satisfy the æsthetic requirements of our highest civilization. In its application it still is a troublesome element of building, and in our endeavor to escape from it, many substitutions have been tried. Of late years we have seen more and more the efforts of wall-board manufacturers to produce a material which will give the desired effect of interior finish without involving the tedious processes that plastering requires. The success of these is problematical, but the effort is a commendable one. It is not impossible that a new standard will be evolved, particularly for moderate priced habitations, but it is doubtful if anything will ever be found as completely satisfying as the hard, true plaster surface that has followed civilization down through the ages.

The use of structural steel has introduced an innovation in the form of plain ceiling panels that architects have been quick to interpret in terms of architectural precedent. The beams of the ancient ceilings jutted down into the rooms and were structurally necessary and accepted for what they were. When lathing was introduced as a refinement of construction, it was done to carry the plaster surface on the soffit of the floor structure above, and architects were prone to cover these beams in the better structures, thus forming the traditional flat, smooth ceilings. When the timbers from above were too deep, they were allowed to obtrude and did no harm architecturally. Now, in modern construction, the steel girder has about the same effect. The construction of each floor is such, particularly where concrete arches are used, that concrete passes down around the beam or girder and back up to the slab, thus forming the ceiling into a series of panels or coffers which prevail throughout the whole structure. This construction is not so marked where hollow tile arches are used because they are generally ten or twelve inches deep, the same depth as the average run of beams of the floor construction, and therefore only the deeper girders protrude. A concrete arch, on the other hand, is only about four or five inches thick, and as this construction requires the same beam depth as a tile arch, the reason for the panels on the soffits of concrete arches is apparent. Generally, where concrete arches are used, plaster is applied directly to the arch, and the occupant of the room or office in which this construction occurs can tell at a glance whether the arch is concrete or tile.

This test is not infallible, however, because some buildings with concrete arches have suspended ceilings under the beams, thus forming a continuous, flat ceiling, excepting where very deep girders may project down into the room. Such extensive use of suspended ceilings is, of course, a refinement only justified by special conditions of the building in which it may occur.

CHAPTER XX

THE DECORATIVE INTERIOR AND FINISH

IF artistic plaster ceilings and walls are the crowning glory of a richly decorated interior, the use of beautiful marbles puts the seal of enduring charm on the artistry of the design. The kinds and varieties of marble available are as many as the pigments on an artist's palette, and where the use of color does not degenerate into extravagance, it is almost certain that a beautiful marble finish will do its part in enriching any decorative composition. The fact is that marbles of themselves are exquisitely beautiful. Like exterior stone, they carry with them the marvels of natural formation, but their beauty does not end there. When highly polished, the depth and interest of the coloring, the very fact that nature herself made them, the interest that the geological structure evokes, and the amazement at the natural phenomenon itself all contribute to the intrinsic beauty of marble as it is used for decoration. It has a fitness, too, in that it proclaims its own wearing qualities and suggests the everlastingly substantial character of the structure. Used by the ancients in their classical buildings where the stones were hewn and brought from the quarries and fashioned with such miraculous beauty, marble has ever held its place; and in our modern skyscrapers, where limited space requires the utilization of every inch, it is natural that the older forms should still prevail, even though the ingenuity of modern marble sawing and cutting has to be brought to bear to enable the material to take its place in the scheme of decoration.

The marble we see in wainscoting is generally about seven-eighths of an inch thick, securely anchored to the partition and held in place by bronze anchors and dowels. Marble floor slabs are seldom less than an inch and a half thick. Plaster of paris is used extensively in wainscot setting, although marble floors must be laid in cement to withstand the shock of foot traffic over them.

We have seen that many of the beautifully colored marbles may not be used for exteriors, but in the interior they hold their place, because they are not exposed to the rigors of climatic changes. This very protection fosters the use of marbles which of themselves are intrinsically shaky and unsound, but when carefully sawed, they are cemented to "linings" of some homogeneous and sound marble and their exposed surfaces are worked and polished to give the effects we see.

The Roman builders knew the art of working unsound marbles and their methods were not unlike our own where interiors were concerned. They mounted marble slabs against their masonry interiors, and it is to be suspected that the system of inspection was not unlike our own. Our word "sincere" comes from the Latin *"sine cera,"* meaning without wax. The word, in fact, refers to the integrity of the old Roman builders, who would not resort to wax in touching up their beautiful marble creations in the hope of getting the work passed by the vigilant critics put over them to see that the marble work was sound and free from cracks.

Marbles used in the interiors of our finest skyscrapers may come from almost any part of the world. The quarries of Greece, Italy and France still furnish vast supplies of this beautiful material. The unlimited variety of color has been touched upon, and yet it may be of interest to note that in America we have some marbles of a beauty and quality found nowhere else in the world. For example, there is no marble that compares for certain purposes with the pink Tennessee marble so freely available to American builders. Many of our Vermont, Tennessee, Georgia and Alabama marbles are close-grained and beautifully figured, and in fact, it may be said that, wherever our mountain ranges have cast up marble deposits, there are sure to be formations of unsurpassed beauty that only await quarrying on a sufficient scale to insure their being used for interior marble work. Architects are ever on the alert for suitable new varieties, and builders are as alert to be sure that the material offered is really available in the quantities and at the time that swift completion will require.

It would be futile to attempt to describe even the commonest marble. It is hoped that the reader's interest will be aroused so that he will inquire, where information is available, as to the kind and origin of the marble finish that he sees.

The process of working interior marble is about the same as for exterior work, excepting, as has been observed, the interior marble is generally cut in thin slabs for all the plain surfaces and with due regard to economy of space in mouldings and bases. Like exterior marble, the work is done almost entirely by machinery, excepting where intricate carving is involved. Mouldings are run on planers, and the polishing of plain surfaces and of running mouldings is almost wholly mechanical. Marble working is an art in itself, a skilled trade, and the artisans that execute the intricate work are selected for their ability and long training.

Mosaic floors and marble mosaics are kindred heritages to decorative marble and perhaps have a more ancient origin. While there has been a diminution of the use of mosaic in late years, due to the high cost, nevertheless it is still the beautiful medium of the expression of sumptuousness in the design of fine interiors.

Perhaps our best relics of decoration from the Roman Empire occur in the excavations of the buried cities. Pompeii in particular has yielded superb examples of this work. It is natural that floors, wherever found, should be the best preserved, as the rubbish of destruction succeeding the downfall of the great civilization in which they were built fell on the floors and protected them from the damage of the elements and of vandalism. So it is that to-day our architects can reconstruct many of the great floor designs of the Roman Empire, where the art of mosaic work reached its greatest height. The Romans laid their floors in "puzalon" cement, a ground volcanic cinder that resembles our Portland cement in its igneous origin, but it never equalled our modern product, either in hardness or uniformity. Sentimentalists like to point to this ancient product as evidence of Roman superiority in building. The material is still used in Italy by the peasants, but is no more to be compared with modern Portland cement than mud bricks are to be compared with our burned brick.

In the memory of builders living to-day, mosaic floors were laid piece by piece, generally by Italian artisans who, with a fine appreciation of the design on which they were working, carried out the intention of the designers, and sometimes actually improved upon it. We can remember these men working on their knees in great floor areas with bags of different colored little cubicles of stone at hand, laying the intricate patterns.

Through the increased cost of labor this has largely been changed. Now the mosaic patterns are almost always mounted on paper in the shop before they are delivered to the building. The process can readily be imagined. The artisans work at benches with paste pot and stout manila paper. The design is indicated in reverse on the paper and the stones pasted on it face down. The pieces of paper are of uniform size, generally about eighteen inches by two feet; or they may be any other convenient size. As the sections are finished, they are carefully numbered and packed in boxes to be sent to the job. Now the workman simply lays the section of paper down on the prepared bed of cement mortar, being sure that it is level and in position, and with flat blocks of wood the stones are tapped down into place, the mortar squeezing up in the interstices between them. The work having been brought approximately to an even bed, the water in the cement, together with water that is poured on the paper from above, soon dissolves the paste and the paper is removed, thus leaving the completed pattern in place. Any small irregularities of position or the adjustment of joints between the sections are now made, and the work is left to harden through the action of the cement setting.

After hardening, the surface is ground down, generally by grinding machines, although we still see the ponderous scraping back and forth of a fragmented grindstone set in a hoe-like implement, which reminds us of the tedium of the labor that built those ancient floors in the Roman Empire.

Wall mosaics are similarly produced on paper mountings, and even mosaic ceilings, now seldom seen, can be done by this offset method. However, with ceilings the problem is likely to be more difficult as, on the rare occasions when we see marble mosaic ceilings, they are apt to be vaulted or domed, in which case the old hand method of setting piece by piece is probably the most economical.

Consideration of marble mosaic floors recalls the fact that one of the survivals that has carried through the changes of style and technique from those same early Roman times is the terrazzo floor. It seems never to lose favor and is properly one of the great, enduring floor surfaces. Unlike mosaic, particles of marble are "sown" into the bed of plastic cement mortar, much as a farmer sows grain; but the work, nevertheless, requires a skill and technic that still commands the interest of some of our best Italian artisans. There is, of course, endless variety of combinations of marble mosaic and terrazzo, or terrazzo itself, done either in plain color or in patterns, although, when terrazzo patterns are used, they must be of simple formation within the possibilities of the rather cruder technic of laying.

Terrazzo suffered a somewhat just criticism, after its wide

use through the period succeeding the early development of the skyscraper, from being used in too large areas. Any floor of Portland cement base laid in a large area is sure to show cracks. Our early architects forgot that while they were requiring the scoring of cement floors to provide inconspicuous lines of cleavage for the contingency of expansion and contraction, in terrazzo they sometimes called for large and continuous areas. About a year after completion, these terrazzo floors were sure to show jagged and irregular cracks, generally over girders, because the girder represented a line of weakness in the overlay which formed the bed of the floor; also because of the girders' excessive rigidity in the very scheme of skeleton construction. Thus, under the strain of vibration which all skyscrapers undergo, terrazzo floors, or in fact, any Portland cement base floor, would yield at its weakened points; hence the jagged and irregular cracks. In terrazzo it is more noticeable because of the perfection of the surface its appearance demands.

More recently this tendency to crack in large areas has been skilfully met by the introduction of brass stripping dividing the terrazzo up into sections of modest proportions. One concern that has specialized in laying terrazzo has devised the ingenious name of "Cloisonné" for the work of this kind that it does. The strips are just what they seem to be—thin sheets of brass set on edge deep into the bed of the terrazzo work, laid in advance of the sowing, and in fact, before the plastic mortar bed has been laid. These strips, set true and carefully anchored, form an ideal ground for levelling and truing the cement, and when the process of grinding sets in, the brass is ground off with the other material forming the floor and we get the pleasing effect of the thin metal lines defining the pattern of the floor. When shrinkage cracks occur, they follow the lines of the strips and are not noticed.

The grains of terrazzo are of marble, generally white but

sometimes vari-colored, for practically all of the hard marbles are usable in the making of this excellent floor material. Moreover, the making of terrazzo provides a valuable outlet for the waste of the marble yard. Hard, colored marbles are not so frequently worked, but for the same reason, they are not so frequently demanded in terrazzo designs.

The design having been sown with brass strips now in place, the cement and marble-chip aggregate is allowed to harden, and this floor, like the marble mosaic floor, after a couple of days of setting, is ground down, either by machine or by tedious process of hand grinding.

Scagliola is a decorative material which might be loosely described as between plaster and marble. It is a combination of Keene's cement—a pure white and very hard product of calcined gypsum and alum—and mineral pigments. In making scagliola, finely powdered Keene's cement, looking so much like plaster of paris that it cannot be distinguished, is mixed with water and spread on a sheet of oilcloth in a thin paste. The artisan, who is in fact an artist, has his pigments ready in pans, mixed in a thin batter of this same Keene's cement. He also has vari-colored skeins of silk threads, brushes and other implements of his work, and these he dips in his color mixture and skilfully draws the pigmented material through the thin, wet mixture on the oilcloth. The materials merge in the veining and design that his experienced hands direct, the thin pigments sinking through to the oilcloth, forming the figured surface that it is the intention to produce. For example, if he is imitating Sienna marble, the reds and orange colors will streak through the design, the silk threads being left in the composition to trace the lines of the veining. Or he may be imitating an Alps Green marble, or a Black and Gold, in which case the basic Keene's cement will be pigmented, the white and golden colors to be streaked through the darker material. The Keene's cement now commences to

set, and as it is left undisturbed on the oilcloth, it becomes
tough and yet remains plastic. The artisan must move swiftly
and decisively, for the setting acts rather quickly and all of the
pigmentation of a single section must be done while the colored mixture is still liquid enough to sink through the basic
mixture, at the same time not being so liquid as to run and
blur the design.

The coloration being now built up in a thin paste about an
eighth of an inch thick on the oilcloth, the mass is of such a
tough consistency that the oilcloth supporting it may be handled if care is used. It will be seen that the design has been
constructed in reverse, the part next to the oilcloth being the
surface we eventually see.

Let us suppose that the material is intended for a great,
free-standing column in a banking room. The column has
been prepared in the rounded form with the usual fire-proof-
ing, on top of which a coating of Keene's cement plaster has
brought it to the exact dimensions the design called for, less
an allowance of about an eighth of an inch all around for the
surfacing about to be applied This column is perfectly shaped
but roughened on the outside. Now the workmen, with everything in readiness, pick up the sheet of oilcloth with the tough,
plastic material clinging to it, just as half a dozen men might
carefully raise a wet blanket. The sheet may be as large as
eight or nine feet square, and the men acting in unison raise it
and wrap it around the prepared column. The dimensions
have, of course, been accurately laid out, and the sheet therefore exactly envelops the column with a lap of perhaps half
an inch in the plastic material. Working swiftly, the men
press the sheet into place with their hands, the reverse side of
the oilcloth yielding ideally to the work, their trained hands
knowing by touch just how to work the mass without deform-
ing the now hardening, plastic material underneath the oilcloth. This work done, the oilcloth is peeled off and the sca-

gliola surface is revealed, still somewhat soft and dull in appearance, but indicating the success or failure of the artist to imitate the marble he has copied. If the column be large, the operation is repeated until the surface is completely covered. As each section is laid, the artisan must put on the finishing touch, for where the joints come, a little adjusting, patching and reconciling is always necessary. So while the material may yet be worked, he swiftly attends to this last detail and the column is left to set.

A few days later, the column surface, now set to flinty hardness, is attacked with pumice stone and water, the first step toward polishing. Smaller irregularities having been ground off, the beauty of the final design is revealed for the first time, and by another day's work, the polishers have completed the task, which, if successfully done, appears and feels like a magnificent monolith of the imitated marble.

Scaghola is generally so skilfully wrought that even an expert has to look closely to discern it. Practical builders, of course, know that a solid marble column, such as it appears to be, could not possibly have been constructed in such a place, nor would such a column stand the excessive weight from the superimposed loading of the upper stories; only structural steel would stand such a strain. But the amateur may always discern the work, however perfect the imitation, by examining the column all around. Somewhere, the patching of the joint will reveal itself, but to no discredit of the men who did the work. Joints are placed, if possible, on the less conspicuous side of the column; or if the column be so located as to be equally conspicuous from all points of view, there is no escape, for the joints will show. And the observer may still rejoice in the perfection of one of the most exquisite imitations done by man's handiwork that the art of building affords.

It will not be necessary to describe here the simpler operations of scaghola for wall surfaces and for pilasters. The initial processes are the same, but the whole work can be done at the bench, the pieces being set in place and finished much as is done with natural marble.

Marble, mosaic and terrazzo in one group and tile in another are generally put together by builders in considering their time schedules and plans of operation. They come at about the same time in the interior finish, and while the extent of either group prevents even generalization as to their place and importance, both are scheduled after plastering, and count as final operations.

Tile has a very respectable antiquity as a wall finish and for floors. The Italians of the Middle Ages brought it to high perfection and, as we have seen, Della Robbia put the seal of the highest artistry on this ceramic medium in which he produced those marvellous masterpieces that have come down to us. The Spaniards and the Moors have contributed immensely to the art. "Spanish tile" flashes to the mind's eye, the gorgeous, flamboyant exteriors of the tropics; "Moorish tile," the splendors of the mosques of Algiers and of the Alhambra. True, the terra-cotta manufacturers claim Della Robbia, but after all, interior decorative tile is only a refined form of the architectural terra-cotta we know. Like terra-cotta, ceramic tile has received a tremendous impetus from our development of the skyscraper. The building of sumptuous hotels and apartment houses has given momentum to the development, for only through the receiving of orders for large quantities could the manufacturers justify the research that has brought about the diversity and perfection in tile that we take as such a matter of course.

Wall tile is somewhat different from floor tile in that the slip or finish is the goal, without regard to the hard wear that the floor tile must withstand. Also the slip or enamel gives opportunity for wide variation of color and texture and, of course, lends itself to ceramic design even to the point of fine china painting. Relief work and the polychroming thereof are only a phase of this beautiful art.

The tile setter follows the plasterer in the building and is a law unto himself in the space in which he works. The walls are laid, true and straight, to the base course, which is first set for the entire room or surface under work; then, course by course, the wall is built up, much as a bricklayer lays a carefully planned exterior, but here the accuracy must be almost absolute. The perfection of the material demands perfection of workmanship; the finished surface is a thing of beauty and the observer is properly critical.

After the walls come the floors, laid on a bed of Portland cement; the borders, if there be any, first laid out and accurately fitted. Then the field, now almost universally mounted on manila paper, as has been described for mosaic, excepting that the pasting on is done by mechanical means, so completely standardized has this aspect of the art become. Tiling must not always be thought of in terms of sanitary work, although a vast proportion of our national product is devoted to that department of building. More and more we are seeing handsome tile interiors, complete decorative effects, walls, floors and ceilings. The famous Della Robbia Room in the Vanderbilt Hotel in New York is an example, which, by the way, includes one of the most exquisite examples of Guastavino arch and vault work in existence. We have spoken of Guastavino tile in connection with floor construction. Beautiful tile is a worthy rival of marble as an enduring decorative material. It imitates nothing and is a decoration by itself through its own, sheer, intrinsic beauty.

CHAPTER XXI

MECHANICAL INSTALLATION

BUILDERS group plumbing, heating, ventilating, electrical work, elevators and even refrigeration, when it occurs, under the term "mechanical installation." There may seem little that is mechanical about some of this work, yet the classification is particularly apt, for these branches must all be considered together, and as a whole involve the use of a considerable amount of machinery. They interconnect in one way and another, and the whole mechanical installation in a skyscraper bears a remote analogy to the circulatory and nerve system of the human body. Without these facilities, the skyscraper is an inert corpse, useless and dead. It would be impossible in a popular outline such as this, to attempt any technical description of even the most general phases of the mechanical work. Rather, it is the intention to point out the existence of the salient features that the reader may glimpse the complexities here involved. We have seen that engineers, collaborating with the architect, have designed the mechanical installation, just as structural engineers have designed the steel skeleton. A few architects essay to do all of this work under their direct, personal supervision, but their efforts are likely to be analogous to those of a physician who would undertake to be a specialist in surgery, nervous diseases and perhaps a half dozen other ailments now generally conceded to be best performed by individual specialists.

The mechanical engineer makes mechanical drawings with the mechanical features, pipes, conduits, motorized equipment, etc, emphasized. These drawings differ both in appearance and purpose from the architectural drawings, although of course both must agree in their ultimate intent. Before he can begin his work, the mechanical engineer must study the survey of the site which indicates the location, pitch and depth of the adjacent sewers, their capacities and the existing demands upon them. It sometimes happens in moderate-sized cities that sewer enlargements must be made, owing to the added demands for capacity that the proposed skyscraper will make. In like manner, the size, location and capacity of the water mains that are to serve the building must be examined, while at the same time, the available supply of electric current and its characteristics must be known and considered.

The mechanical layout is generally the work of an organization that includes specialists in these lines, and while they work as a team, the staff members take up their several specialties and, co-ordinating them, design the complete, interlocked and inter-dependent mechanical system. For example, these engineers use certain formulæ, on which radiation and the supplying of steam thereto, with the incidental pipe sizes, are accurately worked out. The architect must be informed of the size and location of the radiators, and these may differ in different climatic zones of the country. When it comes to heating and ventilating entrance lobbies and large banking rooms, the problem becomes exceedingly technical and complex. Heating and ventilation are here coupled, and of recent years, air conditioning has come to the front as a new and now almost indispensable requirement in certain cases. Air conditioning means that not only must the air be heated, but it must be washed and dehumidified, and even chilled in summer. Patrons of moving-picture theatres rejoice in this recently developed luxury, and there are now few large cities that do not boast at least one splendid new auditorium where the air on a hot summer day is as cool and refreshing as at some seaside resort.

The boiler room deserves early consideration, and the architect must be informed by the engineers as to its depth and size, for the first operation of construction is affected thereby; the excavation must be made to accommodate the boilers. Here again the structural engineer is affected, for the deep boiler room means extending the columns down to lower foundations, and here the foundation design must take this depth into account. Or the plumbing layout must be based on the number of toilet rooms, wash basins and water-using facilities generally, and here the boiler requirements are again to be considered in estimating the total water demands of the building. So the plumber must know what the steam-fitter is doing, for the pipe sizes are affected and must be large enough to handle these loads.

If the boilers are used for heating purposes only, they are spoken of as "low pressure" boilers; that is, they will carry only a few pounds of steam pressure and are not intended to develop the power necessary to drive any pumps or machinery. The boiler feed pumps, the house pumps and the fire pumps in this case are electrically driven, and the electrical engineer must know what the pumping demands may be; also he must know the ventilating and air conditioning problem, to provide the electric feeders and specify the power loads so that the electric motors for all this apparatus may be of proper size.

So it goes, endless interdependence between the mechanical branches, a hopeless complexity to the layman, but relatively simple to the experienced engineer if he but use his head and keep himself posted as to the ever advancing improvement in the art of modern building. The spectator from the curbstone sees the steel erection carried swiftly skyward, and if he but look a second time, he will see that right behind the steel, almost simultaneously with it, the pipes are carried up hugging the columns and securely anchored at certain floors on the beams or girders. It may not be as apparent to him that these pipes often hug into the depths of the column section, and it may be of interest to know that the position of these pipes has been laid out by the engineer for the architect's guidance and in collaboration with him quite as accurately as the structural steel itself. Plumber and steam-fitter are here right on the job with their mains, with all the fittings accurately installed to take care of the branches that will follow. Also, the electrician is here, and if the floor arches are of concrete, his men are swarming over the forms to install their accurately-placed outlets and conduits before the reinforced concrete is placed. Somewhere back in the building the pipe shafts and conduit shafts are tucked away, and in these locations the workmen installing the mechanical work are busily putting in their main feeders and risers, for it is easier to work in a shaft location before the partitions are built than in a shaft after they are built.

The steel has not risen a dozen floors when the elevator constructors arrive with their elevator guides. These men are concerned in the plumbness of the elevator shafts, and watch the plumbing of the structural steel as it goes up above them in the expectation that their work, started perhaps before the steel is half finished, will continue to give them an unobstructed, absolutely plumb, vertical rise from where they are working to the very top.

Sheet metal workers are watching for their chance, which, as has been seen, comes soon after the derricks have left the ground floor. At this point, the stair-builders appear with their iron stairs, for they too can work better on the steel before the masonry walls and partitions are started; also it is important to the builder to have these stairs set, and he quickly provides wooden treads so that the job may have established highways of foot travel up and down.

The builder has determined from the plans the location of his material hoists and these are placed so as to cause the least possible interference with the work in general; but more particularly, these hoists must not too greatly interfere with the mechanical installations. Hoists rise unceremoniously right through the middle of rooms, the floor arches frankly omitted where continuity would seem to demand them most. Hoists are seldom found in hallways, corridors or stair wells, for the reason that in those positions, they would interfere with essential structural operations. For like reason, the elevator shafts are left strictly alone to give unobstructed progress to the elevator constructors at their important work. Permanent elevators are one of the major goals in the builder's programme, for until these are ready, the temporary hoists must remain. It sometimes happens that it is desirable to erect a temporary hoistway completely outside the building to accommodate a swift construction programme, but it is apt to be a costly procedure and is resorted to by the builder only in extremities.

The mechanical installation is timed so that the water tanks and heavy apparatus destined for the upper parts of the building shall arrive when the derricks are working on the floor on which these facilities are to be placed. When the top is reached, there is generally a number of bulky pieces, such as fans, fan motors, elevator motors and like accessories, to be set, and the steel derrick handles these into place. Equipment of this sort is more or less delicate, and the protection and care of it until the building shall be enclosed become a part of the problem.

Is it a wonder that the job superintendent is apt to be found in his office at any hour of the night, or that he is only to be found there at odd times during the day? It is a strenuous and exciting business, but also an inspiring business, for nowhere in man's endeavor do we find the elation that attends a swift-moving, well-organized skyscraper operation. The thrill is always there, the unexpected is always happening; the satisfaction of planning in the welter of all this activity, and of having the plans come out right, of seeing the beautifully finished building come true and clean out of a complexity of elements that only a trained builder understands, this is in itself an unparalleled triumph that gives a man the satisfaction of knowing that he is proficient in leadership.

CHAPTER XXII

LABOR AND BUILDING

THE building problem may be said to be, in a large measure, the labor problem. In the mind of the public, labor seems to be the whole problem. No large building project is ever discussed without an anxious inquiry about labor, and it is all too true that the everlasting bickerings, strikes and turmoil of building labor furnish that unreliable element that has led the public to regard a building operation as the most hazardous and uncertain of business undertakings. It might be ventured at this point that all of this uncertainty is not chargeable to labor, although labor has much of it to answer for. The owner, through his lack of understanding or withholding of decision, furnishes some of it. The architect, through trying to straddle between decision and indecision, furnishes a certain amount, and the very nature of the complexity of building furnishes a share that is never adequately considered.

A factory owner who turns out a product year after year in the comfortable surroundings of a modern plant can tell to a nicety just what that product is costing. When he comes to his building, he inquires the number of thousand of brick, the cost per thousand of laying them, and then proceeds to figure for himself the cost of his work. Not he nor any man living can tell in detail how a stormy week sometime in the course of the construction will affect the cost of the brickwork. In the first place, that particular brickwork is unlike any that was ever done before, a new creation in an office building, let us suppose, designed and constructed for a particular situation that has never been experienced in exactly the same way. True, certain standards prevail, but as a creation, the building is a recombination of a thousand and one elements interspersed with new ideas and perhaps the owner's peculiarities of requirement—and the unit price of laying a brick, the owner thinks should be known to a nicety.

And building labor, in the turmoil of its stormy, belligerent existence, has never had time to stop and think of these things, much less has it been of clear enough conscience to permit sober discussion of the subject. Finally, it has concerned itself not a whit with giving—the hard rules of necessity of employment take care of that—but it has bent its whole energy to receiving. "How much money can I receive, and what power can I exert to compel the giving?" has filled the horizon of building labor throughout its modern, unionized existence. No wonder that unionism strikes consternation in the mind of the uninitiated owner who has read in the newspapers year after year of the brawling and disorders of violent building trades strikes; the bombings, the "entertainment committees," the violence and property damage.

Yet there is another side which no builder can in justice overlook. Intermittency stalks as a spectre throughout the building trades mechanic's life. He has no fixed employer, but must needs travel from job to job performing his highly specialized task in the narrow confines of his trade's unionism, set by the customs of the trade, aggravated by an even greater artificial limitation that the manifold unions connected with building impose. An employer must, by the very nature of his relationship, snatch the man up, use him for a few weeks, and incontinently lay him off on the subdivision of an hour. If he works for the same employer again, it is likely to be on another job, under another superintendent, as though it were a wholly different employer—and again the length of employment is uncertain. Inclement weather breaks up his time. Errors of management that interrupt the smooth flow of material, delays of the work wholly unexplained to him, all militate against continuity of employment; and his detachment from personal contact with his employer leaves him little human relationship on which a mutual understanding could be discussed. A few, a very few, favorably known workmen receive measurably steady employment and are known to their principal employer, but their number in the great mass of metropolitan building trades artisans is negligible.

There is a tragic pity about all of this, for there is no thrill dearer to a builder's heart than the thrill of knowing and talking with and being on terms of friendly intimacy with the artisans doing his work. They have so much in common in the pride of accomplishment; some splendid utility created that both can discuss and admire in the spirit of common understanding. Unlike a machine operation in a factory, building is one succession of new and stirring incidents, always something fresh, always something finished and behind us, with new adventure beckoning.

If the builder has endeavored to translate the skill and ability of his men into dollars for himself without due regard to their equity in the accomplishment, he has at least been over-accused of avarice; for the builder is fortunate, and at the same time possesses a high degree of business acumen, who can, year after year, keep his overhead organization efficient and well employed, and make a profitable income commensurate with his effort and risk. So we have the picture of two factions arrayed in opposition through the very nature of their common occupation.

Labor leadership does little to close the gap. We have much peace-time vaporing about the community of interest, but only in the last few years have discernible traces of the substance of such interest been in evidence. Neither does business turn its hand. The building industry, disjointed, disorganized, with a clientele suspicious and largely uninformed of its complexity, with an architectural profession almost equally uninformed

and clamoring for a recognition of superior knowledge of the
problem which it never possessed and cannot maintain, with the
banking and lending institutions throughout the country taking no stand for a stabilized industry, but relying on an assumed satisfaction with plans and specifications made in a
medium they do not comprehend and written in a technical
language that they cannot fully understand, with bonding
companies as ready to insure the performance of an inexperienced beginner as an experienced builder, so long as the premium is paid, with importunate novices clamoring that they
can build cheaper than any one else, with the sheriff waiting
in the treasurer's office while frantic collections are being gar-
nered in the banker's office to stave off for another brief
period the hand of bankruptcy that overtakes fifty per cent of
his kind in every five-year cycle—is it any wonder that we have
never met seriously to stabilize labor relationships? Is it not a
greater wonder that absolute anarchy in the industry does not
completely overwhelm it?

The state of mind into which modern labor leadership has
put building trades labor, together with the many immunities
thrown around the unions by law in a time when labor was so
definitely the under-dog, has freed the labor leader from practically all restraint. Enormous power unaccompanied by responsibility either to law or, fiscally, to their own followers,
has had its inevitable result. To me the everlasting wonder is
not that there is occasional graft, but that there is so little, the
opportunities considered.

In some instances these opportunities have invited into labor
a predatory type of Tweed politician, who finds the hunting
better and safer here than in a chastened municipal government that used to be his private preserve. He is no more concerned with labor as such than a bookmaker is in improving
the breed of race horses, and being unburdened with scruples,
he has a vast advantage in the struggle for control with labor

leaders of another type. But fortunately, perhaps providentially, the majority is another type of leader, men with all the unselfish zeal, single-minded devotion to a cause, and personal probity of crusaders. They sometimes are fanatics, but their honesty is as fanatical as their economics. They are awake to and concerned with the situation, but handicapped in combating it by the refusal of the rank and file to see beyond the ends of their noses.

Hostilely arrayed, both types, however, not only are agreed on a policy of securing the highest possible wage for the worker, but manifestly connive at slowing up production. Its purpose is to spread the work over a longer space of time. It manifests itself, for example, in the increasing refusal of carpenters to hang doors that have been mortised for locks at the mill, their refusal to set "assembled" trim; or, in the case of the plasterers, the refusal to set "cast" mouldings, which must be "run" in place, tediously and at much greater expense.

But capital or management being more concerned to-day with production than with pay, its answer has been to take as much thought and planning off the workmen as possible and so to simplify every task that the craftsman has only to lean against it to accomplish it. That is, as labor becomes more and more expensive per unit of work, capital endeavors to buy fewer and fewer units of work to be done at the site. It accomplishes this by lopping off non-essentials, devising new short cuts and by substituting the simplicity of the machine for the complexity of manual labor. Building labor is ever skeptical, and even now many an economy or short-cut of construction is forbidden the builder by union regulation.

Capital's purpose is to keep costs within competitive bounds that it may live; but witness how the process has worked out, unexpectedly, to produce simpler and better building. For example, the general substitution of metal doors and windows for wood in skyscrapers was brought about primarily by labor's dragging its brakes. By fabricating window frames, trim and sash in metal units, we not only have more enduring and fire-proof windows, but we save six, separate, slow operations divided between several crafts. Cement floors not only are similarly superior to wooden floors, but their laying is a much simpler operation, and they can be installed as soon as the arches are ready and the pipes are in, while all woodwork must be protected from the dirt and tear of preliminary operations.

So wholly and powerfully unionized are the building trades that their wars now are waged, not against the employer, but largely among themselves, and the dictionaries must needs revise their definition of the word "strike." Ninety per cent of all strikes in these trades to-day are the outcome of jurisdictional disputes between crafts in which the employer is a helpless bystander and a circumspect neutral.

A quarrel between the bricklayers and plasterers, breaking out in Florida during the boom, spread nationally. Neither would work on a job where the other was employed, and building was so crippled that the Hon. Elihu Root was called in to arbitrate the issue at last. Plastering and bricklaying have been closely allied crafts and it has been customary, outside of the larger cities, for the two unions to exchange working cards, an arrangement advantageous to both and to the employer. The bricklayers suspended this interchange in an effort to swallow the plasterers in Florida. The plasterers replied by endeavoring to swallow the bricklayers; the original issue was subordinated and the struggle became a contest between rival leaders for power.

The introduction of metal window frames and metal trim provoked a quarrel between the sheet metal workers and the carpenters as to which should do the metal trim. The carpenters, though numerically still a very strong body, have been badly crowded out of metropolitan building as less and less

wood has been employed, and reduced principally to temporary work such as arch centres and special scaffolding. They
had always had jurisdiction over windows and trim, and they
did not intend to lose any part of what little remained to them.
The sheet metal workers replied that that was the carpenters'
hard luck; that the trim was metal and therefore obviously
not a carpentry job. This dispute went on for ten years and
was adjudicated repeatedly. Whichever side lost repudiated the
arbitration and took to the field again like a Mexican general.

Builders who operate in the field of skyscrapers nowadays
seldom sign agreements with labor unions. No union agreement stands in the face of a jurisdictional strike, and while a
fantastic logic is sometimes attempted by the contending
unions to show that it was the other fellow who broke the
"agreement," it is perfectly well understood that the employer is simply unfortunately in the line of fire of the com-
batants.

This jurisdictional strike business is, of course, the despair
of the best leaders, who admit that it is the curse and may
eventually be the downfall of the present system of unionism.
I say "present system," for there will be no downfall of organization in its fundamentals. It represents a mighty and fundamental principle of human advancement, essentially necessary
in the scheme of things as civilization is organized. The pity
of it is that the principle has been so freighted down with the
fallacies embedded from the days when labor was indeed
down-trodden and leadership was ignorant and militant. Capital was then largely to blame, and to-day it is still to blame
wherever it shows oppressive hostility. That hostility engenders proper suspicion from labor. Labor knows little about
the technic of management, finance and conservation, and
all of this capital could teach labor if the barrier of distrust
could be removed.

The guerrilla origin of modern labor tactics bears the ancient traits inherited from barbarism, ruthless yet admittedly effective. Nurtured in combat and enforced through militancy, unionism has had little opportunity to learn the graces of civilization, to say nothing of civilization itself. That it has come so far on the path is the surest evidence of its fundamental soundness, and one of its great problems now is the gap that exists in its own ranks between the occasional savage and vicious ignoramus in labor leadership flirting with socialism or any other ism, and the sterling, capable men who hold the high positions in the labor movement of to-day. By the very nature of their faulty structure, it is often necessary for the better leaders to espouse the cause of the most notorious labor crook. It is the age-old picture that history has shown a hundred times, but which, through the swiftness of the drama, we perceive but dimly. The lower branch of labor to-day is still the field army of insurgency, with combat the only thing it knows and a keen appetite for pillage and plunder. In its upper reaches, labor is trying to settle down to orderly civil government with laws and courts and responsibilities. As yet, it has not the power to discipline nor the temerity to disagree with its field generals who still flout the fripperies of orderly government.

Figures are difficult to obtain, but it is fair to say that the average wage of even the most fortunate skilled mechanic, taken over a period of a year, can hardly be more than a half to three-quarters of that year's earning capacity. Intermittency curses building labor, and the mounting hourly wages have sharpened the keenness of the employer to get no high-priced mechanic on the payroll except at the last possible moment, and get him off at the earliest possible moment. Labor insists on an ever-increasing division of labor, and in the large cities this is carried almost to the point of absurdity. For example, there are wood lathers and metal lathers, stone cutters and stone masons and, of course, the well-known separation between plumbers, steam fitters and two or three other varieties of pipe workers, such as installers of sprinkler systems, etc. This last group all deal with commercial pipe, their tools, appliances and technic practically identical, and certainly the intelligence that governs good craftsmanship in one could be easily and simply applied to another. The few abortive attempts that have been put forth by labor unions to meet intermittency have been as annoying as they are impractical, and while employers have tolerated the wholly arbitrary union rules that are forced upon them, the attempts have made no contribution to the solution of this great economic problem.

Old age, illness and unemployment are, of course, the traditional terrors of the wage earner's life, but the high risk of all three in the building trades has only been vaguely sensed by the labor unions and the solution meagrely, if fatuously, attempted. All unions, of course, have a strike benefit sum which will last as long as the money holds out. The strike stipend to members in good standing is only a fractional part of the daily wage during employment. But economically there would be no quarrel with this if it were not for the militancy which frequently supports a strike for some wholly unessential objective, considered in the light of dollars and cents returns. Of course, the strike for jurisdictional predominance is a spectre of increasing moment that shadows the whole building industry, though wholly senseless economically.

But to return to the basic problems: in actual practice, after paying strike benefits, the union treasuries, so far as one is able to observe, do little more than pay extravagant undertakers' bills in the case of death benefits, and the old age problem is barely scratched. One of the unions has set up a home for the aged and indigent, but it remains to be seen whether it will have all of the defects of public institutional care—the most humiliating experience that can come to a man who has been useful and active.

The game itself is a killer. One passing a large metropolitan building during construction is apt to notice the young, virile men, with nonchalant manner, who so confidently go about their tasks. Few people stop to consider these same men after twenty-five or thirty years of this rigorous, exposed life. They are hearty eaters and gulp their food, frequently carried to the job cold, or if bought at the ubiquitous hot-dog stand, it is generally of the fried variety with little thought of the science of dietetics. Their inordinate use of tobacco and small attention to dental hygiene, nowadays recognized as of such importance to middle-aged good health, leave them susceptible to the occupational ailments which their work sometimes engenders. Necessarily inconvenient are the sanitary facilities, and this, although the builder does his utmost to make proper provision, promotes constipation and stasis which usually are met by drug store quackery. The admiring spectator sees young men, but little realizes the shadow that an uncertain future is casting. The experienced builder, however, sees the prematurely aged building mechanic, sometimes a pathetic figure, standing on the sidewalk week after week, in the furtive hope that a job commensurate with his now narrowed abilities is available for him. Unionism seems to have done little or nothing toward the solution of this, the most vital of labor problems.

The provision for death benefits on the part of the union seems to give an almost tragic satisfaction to these men. They have hardly been told to look beyond, and contemplate the value of life insurance and an intelligent system of pensioning. Builders and the building industry generally are themselves the sufferers, but so long as the relationship is continued as a field of combat, there is no common ground for the solution of these questions by a scheme in which both can join.

So far as I know, there are no scientific statistics on these matters. They should be a most vital concern of labor and labor unions. Indeed, it is not too fantastic to think that there is a mighty destiny for labor and labor unions in paths already pointed out by insurance companies. Certainly their militancy will not get them these benefits, and the cruel fact still pursues them that, although hourly wages have mounted higher and higher, even artificially under field generalship, the wage per year of the individual has not grown proportionately.

Restriction of immigration has greatly aided and is no doubt a magnificent temporary boon to organized labor. Up to a few years ago restriction of apprenticeship and new membership in labor unions in large cities became almost a fetish, until the leaders suddenly perceived that their memberships were dying out at alarming rates and that dues were accordingly falling off. Industry was measurably slowed down, and the exultation of the shallow-minded leader that jobs would have to wait their turn for the pleasure of the building trades mechanic soon found its answer in the unexpected lopping off of non-essentials, the switching of facilities from one kind of material to a wholly different kind, or in the omission altogether of many of the improvements that civilization had wrought. This commenced to bring labor to a realization that back of all the nonsense lay great economic forces.

Somewhere out of this welter of confusion and misunderstanding there will be developed between the builders and labor a plan for mitigating intermittency, for increasing the yearly wage rather than the hourly wage, and for bringing about an abandonment of this senseless hostility which exists to a large measure through the flush eras of building construction. Controversy, kept up by labor leaders who have brought their organizations into ascendancy in this great field of intensive metropolitan development, may some day be superseded by co-operation; but not until the sane acceptance of the great economic laws, and more particularly the bringing of labor organizations within the existing laws that all other men have to follow.

No other industry is so exposed to labor abuses as is building. I am avowedly a union sympathizer and would certainly be a union member if I were a craftsman. That is partly because I believe collective bargaining is a great social advance, and partly because the great bulk of the competent craftsmen in the building trades in large cities are union men. Labor costs are a charge against the property and therefore against the owner, life insurance companies, bond houses, investors, renters and the public in general. Labor is not wholly the builder's problem.

On the other hand, the owner cannot be expected to give battle for a principle; inasmuch as he builds only once usually, he would be foolhardy to challenge the unions single-handed, even if he were financially able. No sane builder would touch a contract that did not contain a strike clause protecting him against all delays due to walkouts. And were the owner both able and willing to fight a strike, his underwriters lying behind him with great mortgages dependent upon the earnings of the building would not give him any encouragement in doing so. When labor trouble arises on a job, the owner goes at once to the builder and asks anxiously, "Can't we fix this up?" Or worse yet, "If you don't fix it up, I'll take the contract away from you."

The only possible answer, and one that has been tried with some success in certain cities, is a combination of owner, builder and bondholder with a defense fund of millions and a programme as militant as that of labor. In such a war the power of labor might be broken. It is a two-edged sword, however, and for my part, I want nothing to do with it.

One of the building labor's just complaints and a basis from which it argues for higher wages is the hazard of the occupation. Not all trades are exposed equally to hazards, but it is unfortunately true that the occupation of any workman around a large building operation is more or less dangerous. This fact has been recognized by our legislators, and compulsory compensation insurance is now in force in nearly every state in the Union. Only five states are without some form of compensation laws as this is being written in 1928. They are North and South Carolina, Florida, Mississippi, and Arkansas. Men who are injured in the course of their employment in these states must seek their remedy under the old Employer's Liability Law; that is, they must sue their injury claims out in court if they cannot come to an agreement for compensation with their employers.

All of the other states have laws which make it compulsory for employers to take out compensation insurance. There are four forms by which the state allows an employer to insure; first, to insure in the state fund, if the state operates a fund insurance; second, to insure in a stock company; third, to insure in a mutual association; or fourth, to become himself an insurer—that is, to assume liability for the payment of compensation direct to his employees, in which case he is required to submit to the department administering the compensation law proof of his financial ability to pay compensation, and further is required to deposit with such department securities for the fulfillment of his obligations under the compensation law.

The rates to be paid to the insuring companies, or into an indemnity fund if the employer is self-insuring, are fixed by law and differ in different states. They are based on the amount of payroll applicable to each class of workmen, and the employer is required to segregate the classifications in accordance with the instructions that the several states lay down. For example, in New York, the rate on steel erectors in 1928 was $24 93 per one hundred dollars of payroll, while in New Jersey the same occupation carried a rate of only $9.90. Carpenters were rated at $17.11 in New York, but in New Jersey they carried a rate of $4.50. Masons at the same time were rated at $6.62 per one hundred dollars of payroll in New York, and in New Jersey were rated at $2.70. There is a reason for this apparent discrepancy, arising out of the difference of experience of the two states in the casualties suffered in the past.

Deaths are, of course, not the only basis for compensation. Also, the specific allotments under a given form of accident are different in the several states, which directly bears on the premium paid by the employer. In New York, on a death claim, a widow would be entitled to compensation during her widowhood, and the children would be entitled to compensation until each becomes eighteen years of age. New Jersey provides compensation to dependents for a maximum period of five hundred weeks. For the loss of an eye, New York provides compensation for one hundred and sixty weeks; New Jersey, one hundred weeks. The maximum amount paid in New York for casualties as above recited is $25 a week; in New Jersey, it is $17 a week. New Jersey is cited simply as an example for comparison. Nearly all of the states require rates comparable with New Jersey's; New York requires the highest rates and compels the payment of the highest indemnities.

In New York state, when a man is severely injured, the Compensation Commission has a hearing on the case approximately every six to eight weeks. The hearings are continued during the time that the injured employee is receiving medical treatment. When he is discharged from medical treatment, the employee is examined by the state physicians, who arrive at some opinion as to the physical condition that the man is in after being discharged by the treating physician. If it is recommended by the state that treatment of a special nature be continued, it is so ordered. Several months after the treatment is discontinued, the case is brought before the Commission, at which time all the medical evidence is taken into consideration by the referee in charge of the case, and an award is made for any physical loss which the injured has suffered, and then the case is closed.

If the injured man is so disabled as to be unable to follow his usual occupation, he is instructed to secure whatever work he can and keep a record of the time he works, together with the amount he earns. If his earnings are less than the amount he earned prior to being injured, he is allowed to receive compensation for two-thirds of the difference between the amount he earned previously and the amount he is able to earn after the injury.

It is apparent from the foregoing rates that the picturesque occupation of the steel erector is about as hazardous as it seems to be from the street. In fact, steel erection is the most dangerous occupation in construction and carries with it the greatest number of accidents. The *Bridgeman's Magazine,* the official organ of the union, in an issue a few years ago, made a résumé of the recorded accidents over the preceding twenty years. From an outsider's point of view, the classification of accidental deaths seems entirely arbitrary and directed to no particular purpose, except perhaps to show the manifold ways in which a man may be hurt; but the truth was there, an appalling record of nearly two thousand violent deaths with a gradually increasing membership over the twenty years, that probably averaged fifteen thousand men in membership for the period. To-day, that union membership is said to be about thirty thousand, with two-thirds of these engaged in hazardous, outside-erection work, the remaining third being occupied in shops and on more sheltered, incidental work; for the members of this union also erect the interior, ornamental iron, and in some localities even control the placing of reinforcing rods in reinforced concrete, an occupation entailing small hazard.

But the record of deaths does not tell the story. One reliable observer who had survived twenty years at this calling states that he knows no steel erector who has followed the occupation for five years without having had a more or less serious accident. Another bit of information vouchsafed by an observant walking delegate is that, over a period of ten years, there was an accidental death somewhere in the national membership on an average of one every thirty-three hours of employed time.

In the line of his duties, this delegate visits injured members in the hospitals of New York City. He states that there are always ten to twenty men from his local of about eighteen hundred members in the hospitals of New York, as the result of the current accidents of their occupation, these cases all being serious enough to demand hospital treatment, some, of course, resulting in death, while a considerable number result in permanent injury.

Intermittency in the steel erector's employment accounts for the loss of at least fifty per cent of his time. It is an unusual operation that takes the steel erectors longer than six weeks, and it generally requires from one to four weeks for the erector to find another job. He is lucky if the next one is of four weeks' duration.

This same union endeavors to maintain an old age and disability pension. Members of at least twenty years' good standing are eligible when they become sixty years of age. There are few applicants for such pensions, however, for the union avers that nearly all of the men who stick at the occupation for so long a time meet accidental death or hopeless disability. The old age and disability pensions in this union are merged and considered as a single responsibility; twenty-five dollars a month being paid to proven beneficiaries. Throughout the national union, the total demands on this fund are about one hundred and fifty thousand dollars annually, and this is met by levies on the total membership. Dues in this union run twenty-eight dollars a year, but this is increased to about fifty dollars a year through special assessments, largely required by the pension and disability fund.

And so we have a few of the high spots of the casualties of the most hazardous of the building trades occupations; also we have a grim picture of these men, practically unassisted, grappling in their crude but practical way for a solution of an insoluble problem. Their whole system is built by the sweat of fellow-member assistance; suspicious of business and of their employers, savage in their attacks, unreasoning and stubborn, not to say ignorant, in their economics, they are, withal, intensely humane in their purposes toward each other when casualties do occur. It is a soul-stirring epic and one that should command the most intensive co-operation on the part of all who benefit by construction, and that means all the elements of our national life, for building is of the essence of our fundamental, national progress.

The liquor question is acute in labor circles, as in all others, and since the enactment of prohibition laws, it seems to form one of the principal topics of conversation among the building trades, as it does elsewhere. It is, of course, known that the American Federation of Labor is squarely opposed to the law in its present drastic form, and since the organized building trades represent about nine hundred and fifty thousand of the two and a half million membership of the Federation, it may be assumed that these artisans are practically to a man opposed to the prohibition laws under which the country now is governed.

Some industries report that there has been great improvement in labor morale since the enactment of prohibition and cite the diminution of accidents and Monday absences. No such report could be made for the building industry. Men occasionally came to work drunk on Monday before prohibition, and they now occasionally come to work drunk. Nobody knows how many there were in the pre-Volstead days, nor is there any record of the number under the present régime. As fair an observation as any is that there is no change in the con-

dition and that, so far as any one can observe, the averages before and after are about the same.

One thing is certain regarding building labor in the large
cities—the men get all the liquor they want, when they want
it and as they want it. The complaint is only over the high cost
and the bad quality. Union delegates report that the men obtain it whenever they please. Inquiry addressed to dozens of
men at random throughout the industry evokes the same ready
response; they can get liquor in abundance at any time on
short notice. Most of the men know how to make it, and many
of them do make it as a recreational pastime. They exchange
and compare information concerning the best formulæ and
the best sources. No case has been reported of a man wanting
liquor and not being able to obtain it freely and immediately.

How to go about the solution of all of these labor problems is not so easy. Business and business men might contribute invaluable advice out of their own wide experience, but
such advice has heretofore always been regarded with suspicion by labor. They have preferred to hew their own destiny.
Yet how can business meet them? On what possible grounds?
There are a few cases where proven criminals have been elected
repeatedly to high union offices. Cases are on record where
men still serving prison terms are re-elected, and from behind
prison bars continue their leadership. Vandalism and sabotage
are too often applauded or smugly blinked by a shrewd union
officialdom. On what possible ground can orderly business procedure meet such conditions?

And the further tragedy of this seemingly hopeless muddle
is that the individual men are generally so sterlingly honest at
heart. Savage and aloof toward any patronizing attitude, they
mellow instantly to the human touch. Discuss baseball, fish-
ing, prize fights or local politics with them and they respond
instantly. Even more alertly do they respond to serious com-
ment on their own craft, in which they almost universally

take pride. The question as to what business shall do in extending the hand of fellowship to these beloved malcontents in the solution of their most acute problems in which they continue to flounder is still almost wholly unsolved.

Co-operation between business and organized labor in the building trades will come some day, and, in coming, it will bring inestimable human benefits. The menace to organized labor is the exploiting union politician and the disdain of business methods and legal restraint which he continually fosters. He panders to the worst characteristic of human nature, and uses his ever-ready, sure-fire slogan by which men, usually level-headed, may be stampeded into continuing folly. Perhaps there is something in the going relationship, now unperceived, that will develop into the medium of a common ground for a programme of mutual assistance.

CHAPTER XXIII

MODERN BUILDING-TRADES APPRENTICESHIP

APPRENTICESHIP in modern skyscraper construction has been woefully neglected by the employers, and the commendable, if random and scattering, efforts of the labor unions to meet the perplexing problem have been aimed as much at creating a source of membership as at the inculcation of knowledge of craftsmanship. The problem is as old as craftsmanship itself, and throughout the ages, the history and development of handicraft have been interwoven with the apprentice problem, and torn between the contentions of the guilds and their successors, the labor unions, on the one hand, and the advocates of free and untrammeled rights of the individual on the other. Throughout the Middle Ages, and in fact down to the beginning of the last century, the problem was a relatively simple one. Even the advent of the early crude machinery that heralded our mechanical age did not change the ancient customs of the guilds in the apprenticing of pupils, which were seasoned by generations of usage and sanctioned by fundamental laws that were almost universally accepted as defining the obligations and rights of apprentices. Then, in the swift upheaval that turned nearly all economic production to a vastly greater and more complicated scale, the machine age, coupled with the tremendous strides of public education, engulfed the apprentice. He was almost lost sight of, and in the two or three generations that followed, the enormously widened horizon of choice of occupation brought with it a thousand diversities of high specialization and opened undreamed-of fields for quick learning. The ultimate of this specialization is now typified by our automatic machine tender who, day after day and

year after year, performs his whole task by monotonously pulling a lever or stamping on a treadle.

The building industry has not been without this trend to
specialization, and one of its bugbears to-day is the tendency to
create quickly trained workmen to do exclusively a small
operation in what was once an incident to a hard-learned trade.

Take the carpenter, whose traditional apprenticeship was
from four to six years. An apprentice of old, with such a training under the tutelage of a master workman, could do all
kinds of carpentry, and indeed, was often a skilled cabinet
maker. To-day we have carpenters who do nothing but work
on concrete forms, sawing and nailing year in and year out.
Such a man can be taught to be skilful in this specialty in a
few months. Or take a carpenter who specializes in floor-laying. Quantity production in so many thousand "squares" of
floor laid is all that is asked of him. Cynics say of him that his
sole requirement is to be strong in the back and weak in the
head, yet the floor-laying specialist is with us—a product of
our high division of labor.

Then there are the wholly new occupations brought into
being by the development of modern materials and appliances
that have no counterpart in, or even resemblance to, things that
went before. Floor-laying by specialized carpenters had hardly
settled itself when the cement floor swept in as a competitor.
A wholly different craft, a relative of masonry, and the occupation as an occupation was hurled into the maelstrom of
union jurisdictional strife between two branches of masonry.
As an occupation, floor-laying was almost wholly lost to the
carpenter. Linoleum as a floor covering has made tremendous
advances in the past decade. Cement floors, come to stay, afford a hard, unyielding surface that linoleum ideally supplements. What trade shall lay this new material? It takes skilful handling, yet the technic of laying it is hardly established.

Consider the steel erector. What apprenticeship shall he undergo? Admittedly, his is a more or less skilled occupation, yet the principal requirement of this spectacular calling is a steady nerve and ability to work on precarious footing at great height. True, he must be something of a rigger, and for these requirements, a sailor's training is perhaps the path to apprenticeship; but it is a long cry from the sea to the busy metropolitan construction job. Sailors who learn their trade are apt to want to stick to the sea—certainly there can be little hope of any fixed scheme of contact between these widely separated callings.

It is to be said to the everlasting credit of the labor unions that they have not avoided their responsibilities in regard to apprenticing, and it is beside the point to argue that they have carried on the work simply with a view to maintaining their membership. Indeed, why shouldn't they, and in so doing, why shouldn't they receive the very highest commendation for having rendered an inestimable social service?

In recent years, the problem of apprentices has been receiving the attention of the building industry, or so much of it as will ally itself in any form of organization that can bring about a concerted action. The late Bert L. Fenner, of the celebrated architectural firm of McKim, Mead & White, devoted a great deal of his time during the last few years of his life to this important question, and through his interest, the New York Building Congress fostered an apprenticeship plan that has become so well established as to warrant the prediction that it will be followed in the large centres all over the country.

The New York Building Congress is an association composed of architects, builders, realtors, property owners, labor representatives, and others whose interest in any way attaches to the building industry. It is a non-partisan body interested only in matters that by general consent are recognized to be of value and benefit to the industry as a whole and the common concern of all. The New York Building Congress makes itself felt through the Apprenticeship Commission, which in turn co-operates with the New York Board of Education. It is a delicate and heady business that the Apprenticeship Commission handles, and it proceeds with circumspection; but the results it has shown justify the assertion that the plan is a working success.

The Commission works in the following manner: When it appears to a considerable number of responsible people in an industry, say ten or more, that directed apprenticeship training is needed, the Commission sends notice to the leaders in that industry and to labor representatives that an apprenticeship plan seems desirable. Then follows a meeting of as many directly interested as can be brought together. If such a meeting sees the desirability of further action, the plan of action is discussed, and, if possible, certain fundamentals of the objective and the procedure are asserted. Following this action, a committee is appointed, consisting of three members of the labor union of the craft affected, and three from the organization of the employers. This committee then goes to work in earnest and organizes a detailed plan of action to bring about what they consider to be a course of training that will fit an apprentice to be a capable journeyman in the line of work under consideration. The number of boys that may be trained under the plan is carefully considered, for it is held to be prejudicial to the industry to produce more craftsmen in a given trade than will be accounted for by the "separations." The word is here used to designate the men who, for any reason, separate from the trade under consideration, whether from death, illness or the abandonment of that occupation for something else. The plan and curriculum having been agreed upon and the places of meeting of the classes having been arranged, apprentices are invited, through the offices of the Commission, the employers, or labor. An interesting fact is that the expenses of the Commission are defrayed, one-half by the labor union affected, and one-half by the employers.

The curriculum generally extends over a period of from three to four years; the classes are generally conducted at night, and the boys must be employed in the trade during the day. Here variation sets in, according to the specific plan and trade under consideration. Some trades require an alternation of shop and class work; the practical application of some of the crafts can only be done in daylight, yet the "pencil work" may be carried on in night classes. The boys, before entering, have the course and its responsibilities explained to them, and they are shown just how they will progress from year to year if they do their work and remain with the course. The wages that the journeyman in the craft is being paid are discussed and understood. The school having started, accurate records of attendance and progress are kept, and as the boy advances by stages, say six months at a time, he is credited just as students in college are credited. Generally, the apprenticeship is limited to boys between the ages of sixteen and twenty-one. A fundamental of the plan is that the boy's wages while at his apprenticeship are fixed, and as he advances they are automatically increased. Thus, a plasterer-apprentice gets $2.40 a day for the first year, $3.20 the second, $5.04 the third, and $8.00 the fourth year. Other trades of comparable ultimate journeyman's wages start and advance with similar wage arrangements.

The Commission, as organized in the year 1928, had an expense budget of about sixteen thousand dollars annually—a surprisingly small sum when the vast amount of work it does is considered. The administration by the Commission appears to cost about four dollars per year per apprentice under training, which indicates that, in that year, about four thousand apprentices were being trained in the nine craft schools already organized. There are three more in course of organization. The classes that were fully operating in the year 1928 and the numbers of boys being trained in each trade were as follows: Bricklaying, 610; carpentry, 1275; marble cutting, 125; granite cutting, 65; upholstering, 105; painting and decorating, 295; electrical workers, 625; plumbing 75; and plastering, 600.

Owing to the diversity of trades and the various stages of progress of the work, it required seventy-seven classes to conduct the training of approximately four thousand boys in this apprentice school.

Here we have the first comprehensive apprenticeship plan since the days of the guilds. It ventures into new fields, in a way, for, unlike the guilds, it must wrestle with the problem of high specialization—the subdivision of the crafts—and it must meet the complex problems of latter-day trade unionism. It is admittedly an experiment, but enough has been done to show that it can be made to work.

Several of the public schools throughout the country are interested, and as has been said, the New York Board of Education is assisting. Inquiry from large cities all over the country is commencing to pour in to the Commission. Builders' associations are commencing to take notice, and, altogether, the great work started in his spare hours by Bert L. Fenner may yet be his greatest monument.

It must not be supposed that organized apprenticeship had been wholly neglected in other parts of the country. In San Francisco, following the years of union turbulence on the Coast that culminated in the formation of an alliance of capital, builders, material dealers, real estate owners and business generally, a plan of opposing unionism such as has been briefly referred to in this volume was put into action, with the result that unionism was practically wiped out in that city. The problem of skilled mechanics became acute, and the citizens' committee having the campaign in charge started schools of building craftsmanship. They reported unexpectedly favorable results and went so far as to announce that plumbers, steam-fitters and like craftsmen could be trained in a few months, bricklayers within a year, and even the difficult trade of all-around plasterer in less than two years. Such were the reports that came to the eastern cities.

Boards of education in many of the large cities have been awake to the problem and some of them have established creditable courses in branches of the building trades. Cleveland, Ohio, leads in this respect, and the completeness of its courses, together with the effective way they are turning out good building mechanics, is a matter of great interest to the building industry.

In the meantime, it may be said that this organized activity is having its effect on the unions and their direct apprentice-ships. Many of the New York building trades have as yet established no contact with the Apprenticeship Commission, yet they are stirring on their own account to see that the gaps left by the "separations" are closed. Altogether there are healthy signs abroad in the land.

Those unions that have not as yet endorsed the plan or brought their activities in line with it, are, of course, continuing the system of apprenticeship or membership replacement that their several problems require.

CHAPTER XXIV

FINDING OUT THE COST

WE have seen the relationship between the owner, architect and builder and their interdependence, if the greatest measure of efficiency is to be obtained in a large building operation. The architect represents the owner in that he is charged
with translating into plans and specifications the owner's requirements, but until the owner shall have been advised as to
the probable cost and availability of a great many things that
may be used in his building, he cannot clearly decide in detail just what he does want. Moreover, a decision to proceed in
a certain way in one line of work may entail a readjustment of
several other lines. Architects sense this complexity from the
very outset and, in an endeavor to short-cut the orderly procedure and further, in their reluctance to yield one iota of
their commanding position in relation to purchase and sub-
contract, they sometimes essay to supply unsupported opinion
to the owner as to costs and things available. There is probably no business of importance in all of American industry that
is so opinion-ridden as construction, and certainly no other
where assumption is built on assumption when it comes to
planning, as in the projection of a complicated metropolitan
building. Moreover, these assumptions as to cost on the part of
architects, and the assurance with which they will sometimes
put forth hearsay and unsupported opinion, is the first step on
the road to the trouble and misunderstanding that at times
seem in such a large measure to beset construction work. True,
some architects know costs and have within their own organizations accurate cost information supported by an organized

312

service which, in effect, is the same that the builder furnishes. Where such a combination exists the architect may be, in fact, a capable builder as well, and no one could possibly object to such a business combination, even though some disgruntled builder might denounce the dual organization as inimical to the rights of his ancient and honorable order.

The architect, as we have seen, is the leader of a group of three—himself, his structural engineer, and his mechanical engineer. Now, modern architecture, of itself, is in fact three businesses, or, to put the matter another way, the architect in his service to the owner must supply three functions which, while interdependent, may be considered independently to illustrate their separation.

First, there is the design of architecture, that fundamental ability on which the profession from time immemorial has been founded. Until the advent of the skyscraper, design was overwhelmingly the major part of the architect's function. The cathedrals illustrate this; primarily they were great things of beauty, their utilitarianism being easily satisfied by their simple, if extensive, floor plans. Palaces and chateaux were similarly simply large numbers of rooms, each with a chimney perhaps, but certainly without a thought of the complexity that later developments have compelled. So the design of architecture came first, and it is because of this that good design in our national architecture has always been looked upon as a prerequisite of popular approval.

Second, comes what may be called the construction of architecture; that is, the understanding of all of the forms of construction and equipment and the intensive application of this knowledge for the benefit of the specific problem in hand. It is here that the engineers serve, bringing to the problem all the appliances and devices afforded by the latest development in the art, always with due regard to the limitations of cost and the extent of the undertaking. The construction of architecture would also include scientific planning and detailing, the obtaining of the most adequate and economical space arrangement and the arrangement of the facilities to the best possible advantage from a utilitarian point of view.

Finally, there is the business of architecture. Here the architect must be a good business man in his relation to his client, in the management of his own affairs, his office and operating forces, and particularly his business relations with the builder and the builder's dependents. It requires a well equipped and organized architect's office to do the thing successfully, for good business administration is as important to the progress of a great building operation as good banking is important to its financing. Some architects are good designers and stop there. They know something about planning and not so much about construction. Again, there are some architects who are both good designers and fairly capable in all matters of planning and construction, but are not so well equipped to handle the business incidental to the complicated affairs in which a skyscraper project can easily be allowed to drift.

However, it is that fortunate combination of well doing in all three branches that makes the modern successful architect's office of to-day. If owners could only comprehend the risk they run in employing an architect simply because he can make a pleasing design, the business of building would be cleared of a deal of the grief that besets it. The pity of it is that right at hand in any of our great metropolitan centres are architectural organizations well-equipped and organized, versed and able in all three of the indispensable branches. Owners are sometimes at fault in their failure to use care and judgment in the selection of an architect, and discover their error all too late. There are cases on record where an enthusiastic owner has succumbed to a pretty picture, conveniently assuming that he must be dealing with a competent architect, without so much as visiting his office or inquiring into his facilities.

On rare occasions the building industry meets that archaic survival that assumes that every one is dishonest, and to him the temptation is very great to foster the fallacy that, were it not for the architect, the owner would be unmercifully cheated by the builder. This outworn fetish has left its scars on the mind of the building profession.

We hear much back-door whispering among builders about how some one knows this or that architect and how, in a certain big job about to be let, a golf course architectural acquaintance is sure to count in his favor. This frame of mind gets the builder to believing that architects give out work and designate contractors. Such is not the case. Architects seldom have more than a negative voice in the scheme of things. The owner who can afford a two-million-dollar skyscraper is likely to be a man who decides his own business deals himself, and while he may have the most complete confidence in his architect, there is not one case in a hundred where the owner turns a large building project over to his architect and bids him go and employ a contractor. What really happens is that the owner inquires at sources of information that he thinks reliable as to the standing and ability of certain builders he is considering. His architect is consulted among others. It is a brave architect indeed who would say that this or that builder is best equipped to do the work. Such a position would lay the architect open to inquiry as to why he advocated any particular one, and while his motive would in no way be questioned, there are not many architects who have any accurate means of evaluating the relative services of three or four leading builders. There may be good and cogent reasons why at a particular time one builder would be better equipped than another, but the standards so far set, both in the building industry and in architecture, preclude anything but generalization on the subject. What the importunate builder is really seeking from the architect is that good opinion shall not be withheld.

One thing is sure, and that is if an architect tells his client not to employ a certain builder, that builder will have an almost insuperable obstacle placed in his way. Off-hand condemnation on the part of some architects is all too common, and in fairness it is to be said that it is doubtful whether architects realize the damaging effect of their negative opinion. More sinister is the architect's condemnation arising from a prior experience, where the builder's unforgivable sin was that he would not accede to the architect's every request. Such adverse opinion from architects is little short of blackmail, and is prevalent enough to be given attention in the standard form of contract endorsed by the American Institute of Architects. In that excellent document, in its later editions, there occur clauses requiring equitable relationship and fair dealing with builders, and one who has followed the development of the industry realizes that these clauses have been inserted in response to a general demand that the arbitrary and capricious attitude that has been scandalously overdone at times, even by architects of standing, shall hereafter be abandoned in favor of equity to all concerned, regardless of the inconvenience that may ensue.

Architects nearly always have representatives on the work, whose function it is to interpret plans, make inspections, and in general supervise in the owner's interest. It is a most necessary service on a large operation, and a capable architect's superintendent renders an invaluable service to the smooth running of the work. No honest builder need fear the architect's man; yet it is important at times that the architect himself do a little supervision of his own people, for the discretionary powers that such a man must have need occasional reviewing. As has been inferred, the great majority of these men are helpful, and, indeed, indispensable factors of progress, but once in a while a builder will have an impossible figure placed on his work; and when this occurs, the work is sure to suffer.

One professional of this latter sort based his claim for consideration for employment on the fact that he had caused the contractor to lose money on every job he had ever superintended. The calamitous aspect of his existence did not strike him. He was intent on proving that he had always compelled the contractor to give the owner more than a dollar's worth for a dollar Such people are, of course, a liability and a scourge on any operation on which they lay their blighting hands. Modern well-organized builders will, of course, not stand for their presence on an operation of importance. If an architect's superintendent cannot perform his function of a helpful consultant and co-ordinator, a medium of quick adjustment and a clearing house of information, he is not worth his salt. Performing these functions well, his office as inspector and critic will automatically take care of itself, and the owner's interest will be more than served by his helpful presence on the job.

One of the problems of both the architect and builder is to acquaint the owner first with what he is going to get, and second, to educate him in the few months of their mutual contact with a technic that has taken both of them years to learn. Building seems so obviously simple to many owners that they are apt to weary at discussion of it. Everybody is sure he knows a lot about building, and there are few men who would not like to try their hands at it; they are so convinced that they could immediately correct the inherent evils and do things so much better than they are being done. The spectator from the street sees the laborer loafing over his task, the plumber's helper smoking a cigarette, and the bricklayer standing idle while a mortar tub is being filled. He knows that he could cure all of these things, and at once he imagines himself a successful builder. The rage of an owner who is paying the bills and sees these things may be imagined, and yet it may surprise him to learn that these happenings are not the principal concern of the builder. True, a builder is alert to correct abuses and has devised a reasonably efficient system for reducing them to a minimum, but after all, this petty loafing, if it is not carried to excess, is but the small-change of an operation. The builder's superintendent is thinking about ways to cut the total number of bricklayers' helpers from three per mechanic to one and a half per mechanic—an arrangement of bins and runways, the location of the hoist with respect to the material storage space, the ways of stocking the floors in advance with brick and tile, the time chosen to do this stocking that he may accomplish his purpose. These are the big decisions and the ones on which the progress and economy of the work at the site depend. Little wonder that the superintendent does not get over-excited about a loafer; his case is easy. The superintendent simply posts his foreman to look out for the man, watch to see if he is caught again, fire him. The thing is done, and perhaps a dollar saved, and it is an important dollar too; but what does it mean if, by not diverting the superintendent's attention to the incident, he is enabled to perfect an ingenious plan of action that reduces the percentage of labor on the whole operation, thus saving thousands of dollars? Owners sometimes dwell on the loafing laborer without sensing the main problems, which can only be solved out of the builder's intimate knowledge of the work and his long experience with similar problems.

Architects and builders sometimes have great difficulty in obtaining decisions from owners, one of the most fruitful causes of delay and increased expense. What architect and builder does not recall the agonizing postponement of decision from Wednesday to Monday, and then to Wednesday again? "Wait for the directors to meet. I'll put it up to them." . . . "Couldn't get the board together; matter has gone over until next week's meeting." These and similar causes of postponement are familiar phrases to almost any builder and architect who has had to do with large buildings. It is not that owners

who do these things lack business sense or enterprise. It is because they are busy and because they fail to understand the
complexity of the problems that modern building on a large
scale presents.

It is for this reason that the real success of a large building
enterprise rests on the absolute co-operation of the owner,
builder and architect. This skyscraper building is wearing
business and experts who understand it and are charged with
the responsibility for it must be given leadership and listened
to seriously, if the best results are to be obtained.

In the public mind, estimating is always connected with
building, and the layman thinks of the builder as essentially
an accurate estimator. Also he is apt to think of the best
builder as the one who can make the lowest estimate, and not
a few builders lay claim to superiority by alleging that they
can "figure closer" than any one else. The fallacies here involved are at the root of many of the ills that beset the
building business, and the misconception on the part of the
layman accounts for about all the troubles he encounters
when he comes to build. Things cost what they cost, and not
what some importunate optimist hopes he can make them cost
by blinking the facts of his problem.

When it comes to estimating, the well-organized builder is
likely to suffer in the eyes of his prospective client because he
insists on weighing all of the elements and contingencies of
a given problem, and then putting a price mark on each element. Hopeful owners do not always like this and are apt to
take refuge from the cold fact by citing what Smith or Jones
did a few years ago. Occasionally it is possible for the builder
to produce the facts surrounding the cases cited and show very
conclusively that conditions were different and the problems
unlike. The chances are that Smith did one thing, Jones another, and the objective of the doubting client is different from
either; yet he thinks of it in terms of something he has heard

of or seen, and his architect's drawings, an unaccustomed
medium of expression, make less of an impression than does
his hopeful memory.

To return to estimating, the quantity survey of all of the
elements entering into a given operation is something of a
science, but the rub comes in pricing the labor element. It has
been observed elsewhere that no building operation is exactly
like another. Every building is specially designed and, in a
way, a specially constructed operation. This may not be literally
true of mill and factory construction, where one unit succeeds
another in accordance with a predetermined plan, but the
statement is true of large metropolitan buildings. Therefore,
not only is the productive effort of labor under conditions that
may arise six months hence rather uncertain, but the specialized thing that the labor is to do introduces a problematical
element as to just how much of the task will be done per hour.

One of the popular misconceptions is that buildings may be
estimated at so much per cubic foot. In the hands of an expert
who understands all of the elements of his problem, it may be
a reasonably accurate method; but in the hands of a layman,
it simply becomes a convenient refuge to which he clings with
desperate tenacity, like a man marooned on a rock in the
midst of an uncharted sea. The use of the cubic foot price by
a builder is analogous to the use of a clinical thermometer by
a physician. It is the first step in diagnosis, yet no physician
diagnoses solely from the thermometer reading. The physician
judges as well by the appearance of the patient, his general
health, and a hundred and one other observations known to
his science. If a clinical thermometer diagnosis were the sure
and infallible means to good health, how simple life would
be! Similarly, in cubic foot pricing, were it at all an accurate
guide, how many of the ills that beset the building industry
would be eliminated—untold thousands of dollars now spent
in careful quantity survey and price study could be saved.

Even when the cubic foot unit is used—and it has taken such hold on the popular imagination that it is doubtful whether it can ever be eliminated—about the first problem to be met is how to arrive at the number of cubic feet. It is a simple matter when we are measuring a box, but it is a very different matter when a structure of irregular form is considered, and particularly where the foundations are of an unusual design. The best rule, and one generally accepted among builders, is to compute the cubic foot contents of a building by an arbitrary line drawn beneath the footings and above any special foundation construction. Where special foundation conditions are to be met, they must be considered as an element of cost apart from any generalization as to cube. From this point up, the extreme dimensions of the structure are considered, and when setbacks occur or courts materially reduce the cubic contents, they are, of course, deducted. When it comes to irregular, sloping roofs, an average well above the exact pyramidal content must be taken to account for the added cost that such roof construction involves. Here again, experience must be the guide. There is one celebrated case where the total cost of the building was based upon its cubic content. In the dispute that arose over the final settlement, the question of how to figure cubage became paramount. The owners, contending for the lowest possible cost, argued that the cubic content of the building was properly the amount that it would displace if it were immersed in water. In their pursuit of this theory, they actually measured all of the depressions of the window reveals, all of the voids behind parapet walls, and finally went so far as to take the diminishing volume of the stepped foundations under the columns and up to the underside of the basement floor.

However, even if a cubic foot price is agreed upon and a contract based upon it, there still remains the necessity for the builder to make an accurate, detailed quantity survey of the operation and an estimate of the material and labor costs. To the inexperienced, this detailed estimating may seem like an interminable job, and it is to be said that there is no royal road to it. With hard, painstaking application on the part of estimators, every square inch of the blueprints is examined, and every line of the specifications read and compared with the drawings. Here individualities of construction and finish crop up—little idiosyncracies of the architect, and the big ones too —and here the special forms of construction and finish and the special material come to the estimator's attention, and he must make note of them and assess a money value against every item. Thus a quantity survey is made of every element of the building from foundation to flag pole.

Now, in this great complexity, there will be from thirty to fifty lines of work that the builder does not do himself. Steel fabrication, elevators, heating and ventilating, plumbing, electrical work, marble and tile, steel trim, and so on almost indefinitely. The task would be well-nigh hopeless were it not for the sub-contractors. The builder has a list of sub-contractors that he knows to be reliable, and it is probable that the architect expresses a wish that on certain lines of work he wants certain sub-contractors that he knows to bid. Cards are sent broadcast through the trades on these lists who are known to supply the various items. The recipients are invited to call at the builder's office, see the plans and submit tenders for their specialties. The builder is organized to give them access to the plans in which they are interested—generally a large room is set apart with plan tables aplenty and reasonable facilities by which the sub-contractors may obtain information. If the builder be one of a number of general contractors invited to submit proposals in accordance with plans and specifications, a day and hour is set by the architect when the general bids will be opened. The builder, in turn, sets a day or two in advance of his own date for the receipt of sub-bids, so that he may have opportunity to compare these and make his own selection of low figures.

In the process just described, the builder will have secured and paid for from fifteen to twenty sets of blueprints, because many of his sub-contractors must do their estimating in their own offices, and moreover, the large number of bidders cannot be accommodated in the builder's office in the time allotted for preparation of the estimate, which generally takes from two to three weeks.

The work the builder will do himself, such as foundations, masonry, carpentry, etc., he is having his own estimators take off while this estimating by sub-contractors is going on, and on the appointed day, with all of the elements of cost fully before him, he tabulates, compares, checks; and the grand total of all the low sub-bids, his own work, his estimated job expense, added together make what the builder is pleased to call his estimated cost.

To the layman, this procedure may look like the most logical and rational that could possibly be devised—an exact science, he might say. But it is not. It is in fact the most haphazard, wasteful, inexact and uneconomical method that could be devised, when the practical working out of the procedure is considered.

In the first place, the builder could not possibly compile his bid without the vast amount of work that is being done by the two or three hundred sub-contractors to whom the invitation to submit sub-proposals has been sent. And in these sub-concerns, the uncertainties of field labor costs, or indeed, manufacturing costs, are the same that beset the builder himself in pricing up his own work. We have seen that all manner of special work is called for, all manner of new designs and special finishes. Each sub-bidder approximates what he thinks his production cost will be for these unprecedented specialties. They all have rules of approximation of their own making, out of their varied experience, but to be able to foretell the exact cost, let me repeat, is impossible. Now when the sub-contractor submits his bid, the builder at least has a fixed price for some element of the building, and to that extent he is in a position to unload on the sub-contractor the hazard of cost variation on that particular item. Fifty sub-lines, comprising eighty per cent of the total cost, are by this process underwritten by the sub-contracting industry generally. The twenty per cent that the builder underwrites again is in part underwritten by vendors of the commodity materials entering into his own work—brick, crushed stone, sand, hollow tile, and like building materials.

Logic seems still to follow, but now again we are confronted with practice that holds little in common with theory. The builder, knowing that all of his sub-bidders are dealing largely in uncertainties of production cost, commences to look the field over for himself. He knows perfectly well that, by adding up the low bids and including his own work and job expense, he will produce the highest bid, not the lowest. For his competitors will do what he perforce must do—go down along the line arbitrarily cutting every low sub-bid he receives anywhere from five to twenty per cent. In other words, he anticipates that, in closing, he will be able to use the leverage of competition in each sub-line to break down the prices of his sub-bidders. Each sub-bidder is nervous lest he be not low, and when he is told by the successful general bidder that there are prices lower than his out for the particular part of the work on which he is bidding, the sub-bidder, uninformed, is talking to the only really informed figure in the whole transaction. One might think that a sense of honor would impel the general bidder to disclose to the sub-bidder that he has fairly won his job; but if that were done by the general bidder, he would go bankrupt, for he has, by his general bid, sold short on every low sub-bid he has received.

Thus we have a picture of the pernicious custom that has grown up in the vicious circle of competitive bidding, a custom that has projected itself into the complexity of modern building from days when vendors of the simple parts of simple buildings sold their supplies to the owner or his representative. Is it any wonder that competent builders will not take part in this ruinous and vicious farce? Is it any wonder that the bankruptcies among builders are fifty per cent of all who engage in it over periods of five years?

The description of this system, or lack of system, might lead the reader to a deep sympathy for the sub-contractor were it not for the fact that the canny sub-contractor knows just how the bidding is going to be run off. He knows that if Smith gets the job, no mercy will be shown, and he bids with caution and reservation to Smith. If Jones gets the work, it will be handled so skilfully that the sub-contractor can afford to lop off five to ten per cent, because Jones, in addition, has good credit. If Brown gets it, the sub-contractor knows that storm signals will go up all along the line. But in the bidding, nobody knows which of the general bidders will take the longest chances in selling short, so the sub-contractors all mark their prices up five to fifteen per cent, put in their bids to all the general bidders alike and await developments. In this free-for-all, the element of uncertainty of actual cost has been worse confounded by the wary bidding of the sub-bidders. It is not unlikely that the architect, keeping open the door to the possibility of himself acting for the owner as a sort of general contractor, will gather in all the sub-bids that he can corral. In this his position is dominant, for only he can modify the plans and specifications, and he generally stands ready to do this to justify his own estimating put forth when he was interesting the owner in architectural services.

The government must of necessity abide by the system, and the result is that an army of camp followers has grown up around government work that stands it in sorry stead when really fine things are to be done. There is no escape for the government, because governments, from time beyond recollection, have been obliged to substitute rules of action for judgment and discretion. It is a rule of government that it is better to squander money on wasteful exactitude than permit discretionary economy. So we have the spectacle of government buildings delayed years in the preparation of plans to the minutest detail and specifications in endless reams of technical and inelastic language. Finally, we see government work started years after it is needed, from plans that are obsolete before the project can be built—ponderous granite exteriors that consume space, interiors equally ponderous in overdone marbles and wasteful of arrangement, the net product of the system by which the work must be produced. Compare the average, "stately" government building, national, state or municipal, with the up-to-date, efficient office buildings that almost always surround them; the stodgy inferiority of the former fairly shrieking from the deadly comparison that the efficient, privately-owned buildings afford. Consider that the product comes about largely from the traditional system of competitive bidding!

Scientific modern estimating in the hands of a capable builder goes beyond the system just described and takes into account certain elements that the competitive system could not possibly have considered. Also, it abhors the fixed, finished plans, and ironclad specifications, for when these have been made, there is little chance for modern estimating. The better way, the estimating of economies coincidentally with the estimating of costs, takes into consideration the market before conditions are fixed. It must be predicated on the builder's being in the owner's confidence, and co-equal with the architect in the decisions of the owner. Here the builder's experience is taken into account and he is consulted. First of all, he must be apprised of that most profound of secrets, the amount the owner expects or desires to spend for his project. In conference with the architect, and furnishing authentic information as to the approximate prices and values involved, the builder is soon enabled to prepare an intelligent budget of probable cost. Every item is considered, and the best and most economical elements in every case having been selected, the architect is informed by the budget and by frequent conference between all three principals to this highly important relationship, the cost limitations of all of the elements he is incorporating.

Now, with the budget estimate established, the builder commences buying on account of the owner. One line at a time is considered and the best sub-contractors in that line are invited in; the plans and specifications are discussed in detail, the owner and architect are parties to the conferences and commence to learn the sub-contractors' estimated cost of doing the thing the way it is specified, and also of doing it the way the sub-contractor is used to doing it. Here the surprises begin to appear. Under any particular sub-contractor's method, exactly the same result can be obtained for less expense—perhaps even a better result. At once, all three concerned in the owner's problem see not only the cost of specialization, but the economy of standardization. Now the owner can decide intelligently just which he would rather have, and he learns the cost of refinements and even of caprices. Only sub-contractors of standing ability may be entrusted with such relationships. They, in fact, become temporarily consultants, but the system brings the inestimable advantage of knowledge of the market to the decision at the time it is made. On the basis of such information, the closing is made, and the sub-contractor is apt to have contributed an economy, at the same time negotiating a basis of performance of his work that he knows will insure him a reasonable profit at least. The budget estimate has been satisfied and safeguarded, and the saving effected here is available as a saving on the total cost or to strengthen some other part of the budget where embellishment perhaps is most desirable.

So it goes. It is not necessary to trace this logical and scientific budgeting and estimating through all of its details. Six weeks spent in intensive conferences between architect, builder and owner in attention to this truly scientific and sensible procedure will produce all of the economies possible or warrantable in a large metropolitan building operation. It has been said elsewhere that the method is not as spectacular as an open, public bidding on an appointed day, but the rewards are rich, not only in producing understanding on the part of all concerned, but also in producing all of the economies that are possible and all that may legitimately be expected. It is here that the truism that buildings cost what they cost is conclusively proven, and it is here that the lowest possible costs are produced.

CHAPTER XXV

JOB ORGANIZATION AND DISCIPLINE

JOB supervision has heretofore been referred to as the activity of the superintendent in marshalling the lines of work into lock-step and, through generalship and planning, bringing the operation to a successful conclusion. Job discipline has been presented in terms of the marshalling of sub-contractors to their several tasks and so managing the intricacies of the work that all of the parts will come together and the building will be the complete and adequate thing that the owner intended it to be. To accomplish all of this there is an internal mechanism known as the job organization, without which there could be little better than chaos as the work proceeds.

The job organization of a large, modern skyscraper operation is a piece of team work for which the builder is responsible, and the measure of a builder's ability may be judged from the skill and efficiency with which that team operates. We see endless truck-loads of material being delivered at all parts of the work, and perhaps wonder what possible system could accurately account for it all in the seemingly distracting confusion that accompanies all trucking in busy city streets. We see high-priced mechanics and workmen of all sorts swarming over the work, and hear of the careful watching of every hour of time, and we wonder as to just what degree of accuracy accompanies this high-priced labor turnover.

In the first place, it must be remembered that sub-contractors account for a considerable number of men on the job, and the time and material accountability for all that they do is a function of the sub-contractor solely. He may have any of a number of systems, according to his general business, or the line of work he is performing. As far as the general contractor is concerned, he takes little account of these sub-contractors' men so long as they observe the general rules of discipline; he is of course concerned with the questions of safety and non-interference with other trades. It not infrequently happens that sub-contractors perform some part of their work, such as extra work done by overtime, changes involving ripping out existing work, etc., in which case, the time and material involved become very much the concern of the builder. When this occurs, the builder's timekeepers and material clerks are charged with the checking and approval, and such supervision then has to be handled as though it were practically a direct responsibility of the builder. Sometimes the extent of this type of sub-let work is such as to require the undivided time of one or more inspectors and checkers from the builder's organization. The sub-contractors keep the time of their own men, pay them off, attend to supplying their materials and facilities, and the job superintendent simply looks for results. He requires that sub-contractors keep the requisite number of men on the work and at times dictates the working of overtime; but with all of these elements satisfied, his accountability ends.

Quite different is the attitude toward the builder's own men and materials. On a large job, there is a principal timekeeper and his assistants; also there is a principal material clerk and his assistants. All materials have been bought under stipulated conditions of delivery, and copies of the contracts covering these purchases are on file at the job. Then there are foremen, the direct lieutenants of the superintendent. There is also generally an assistant superintendent, and the all-important, if inadequately named, job runner, who acts in liaison between the main office, the architect and the sub-contractors.

When the job opens up, a foreman is furnished with blank "Hire" and "Discharge" slips. Let us say a few bricklayers and a few laborers are employed. The foreman of each group indicates this employment on a slip, signs it and hands it to the employee, who is instructed to present it at the timekeeper's window with the hour of employment noted. The timekeeper has a board of numbered brass checks, and upon presentation of the "Hire" slip, notes the name on the slip and in his time book, which is especially made with duplicating sheets, assigns a number which is represented by the brass check which he hands the new employee. The man is now on the payroll and reports to the foreman, and his work commences. This employment is generally done before the beginning of the day's work. The whistle blows, the timekeeper notes the checks that he has given out, and shortly thereafter goes out on the job, time book in hand, and commences checking the men against the numbers, again accurately noting the names.

Working gangs having been built up, and the job in full swing, the process becomes complicated, but nevertheless must be under complete control. Now every morning the timekeeper is at his window a half hour or more before starting time. The men file past in procession calling for their brass checks, the timekeeper scrutinizing each man to assure himself that the name and number correspond. When the whistle blows for starting, the time window is closed. As before, the timekeeper checks his board, noting the 'brass checks taken, also those not taken, which indicate the absentees. This done, he is out on the work again, checking up to see that the men who took checks have gone to their appointed places. A single timekeeper, if he is capable, can in this manner take care of from one hundred and fifty to two hundred men, know them all by sight, get over the job checking the men at least twice a day, and almost instinctively spot absentees without looking at his board. We have spoken of a head timekeeper, for when these big jobs get under way, there are at times as many as five hundred to a thousand men on the builder's payrolls, and in

this case a number of timekeepers are required. Each has a
separate window and is accountable for the records of attendance of an assigned number of men

But the timekeeper has a further function. His especially
constructed time book is so arranged that, as he checks the
men at attendance, he also notes what they are doing, for it is
important to the builder to keep accurate payroll costs, to be
checked against the unit costs of the working estimate. To accomplish this, the builder has designated in a set of printed
rules a system of numeric symbols which the timekeeper must
learn and know. Thus, he comes upon a group of men working on concrete footings, and his attendance mark on that part
of the day is C.1, meaning concrete footings. As he comes to
a gang, let us say, laying common brick, they and their helpers, even to the mortar machine tenders in the basement, are
noted by the M.1 check mark. Now as the payrolls are made
up, these charges are picked out in columns and the totals
show the labor costs in each of the subdivisions of the work.
These go to the cost clerks who, with similar information as
to the material used, make up the cost accounting for the work
for the week.

The material received on the job is checked by material
clerks and no bill for material is honored unless accompanied
by the receipt that these checkers give. The material men, like
the timekeepers, are under supervision, and in spite of the
seeming confusion, there is a receipt given for every load,
whether it be sand, cement, stone, lumber or any other thing
entering into the building. These material men watch for the
sufficiency of loads of commodity material, such as sand, and
are directed to note shortages in the receipts given.

Every week the amount of work accomplished is measured
and reported on a regular form to the general superintendent.
The builder's cost clerks check these reports against the material receipts, for the material weekly reports give the amount

of material on hand as well as the amount used. The watching of these activities and the accountability thereunder demand constant supervision from the main office direct, without passing through the job superintendent, who is busily concerned with progress and planning. Therefore, the treasurer of the builder supervises the accountability direct, and employs the timekeepers and material clerks. These are carefully looked up before they are employed; they are also bonded.

The timekeeper each week sends in his payroll sheets made up to Thursday night for the Saturday payroll. Some of the unions demand pay up to quitting time, and therefore attendance has to be anticipated, involving another review of the payroll at the time it is delivered to the job on Saturday morning. The payroll sheets sent to the treasurer's department must be signed by the superintendent and accompanied by a statement of the petty cash on hand at the job. A certain amount of cash is kept on the job to pay off the men who are discharged, for on a large operation there is always a flow of men coming and going, and the discharge slips, properly signed by the foreman and attached to the receipt received from a discharged employee, accurately account for this cash.

The builder's treasurer has his own checker and inspectors. These move from job to job, drop in on any timekeeper or material clerk, take over and check up on his records as he finds them at the hour of his unannounced coming. This means that these records must at all times be written up to the hour, just as the accounts in a bank must be at all times. Vigilance controls, and all through the accountability and checking are never allowed to grow lax. It is for this reason that we see on the bridge surrounding the building, or later, in the building, these well built, well lighted and well heated temporary offices. It is a large and complicated business that is being transacted in these places; large sums of money are changing hands, and large responsibilities are involved. Only the well-organized and capable builder is equipped to handle it properly, and one of his principal functions is to have a going organization of men who know the builder's methods and systems, men who are trained together as a team.

CHAPTER XXVI

PROBLEMS OF THE FUTURE

THE swift development of the skyscraper and the science of modern construction might be said to differ but little from the equally astounding transition that has taken place in all forms of industry. Modern ingenuity, having visioned the possibilities of vast mechanical accomplishment, is whirling us on toward an inscrutable destiny, and no man can even commence to predict its ultimate goal. All we can say is that the whole effort is toward making human life easier and happier, freeing it from exhausting and inefficient toil, and giving to every one a better balance between leisure and useful occupation. Building, even in its most scientific development, holds to its ancient and traditional course more than does almost any other of the human activities that look to a better and more comfortable world. Destiny beckons us to a future that we feel is to be ever brighter.

In transportation, for example, we have always been ready to abandon the old forms entirely and, without even a backward look, we take to the new, rejoicing. Communication is ever seeking newer and better media, always tolerantly impatient with its latest development. In foods, we accept the endless varieties, and an ever improving knowledge of dietetics may eventually lead us entirely away from an unscientific past.

We have always been ready to abandon with complacency the older forms when some newer form helps annihilate time and space. But in human habitation the requirement and the goal of our ideal has remained rather constant through the ages. Rooms have always been rooms and, as a matter of fact, the ideal of their size has changed but little. Shelter from the

elements almost entirely fulfilled the earliest human requirement, and that same shelter, with the modern additions of
window glass, controlled heat and night illumination, has almost completely satisfied us with the space in which we dwell.
True, sanitation and its corollary, abundant fresh water, with
the incidental ability to choose its temperature at will, are
somewhat modern accomplishments so far as the masses are
concerned; yet even all of these things were early perceived by
humanity as the ultimate of its desires, and these desires have
not changed excepting in the measure of refinement and the
ease with which they are achieved.

What man has done in his building has been to travel in a
great circle of evolutionary detail from the communal cave or
hut out to the separated family abode, then to the further refinement of the multi-chambered domicile. Then, with the
advent of our mechanical age, the tendency has been a return toward communal living, not as a measure of self-preservation such as prompted the earliest communal life, but now
as a matter of mutual self-benefit in the attainment of the comforts, conveniences and indeed the luxuries of life that modern
urban existence offers in such abundance. Hence our cities;
hence our congestion, for convenience of proximity to the
sources and origins of these comforts becomes as important as
the existence of the comforts themselves. We have seen the
swiftness of acceptance of multi-storied structures as soon as
the means of producing them were invented. We have seen
the enormous wealth their invention and the consequent requirements created. But the basic requirement has remained
the same: safe, comfortable, adequate, sanitary and hygienic
habitation of about the same dimensions as originally conceived, and certainly about the same objective of conveniences
to which primitive man first aspired.

Such retrospection, while it can do little in supplying a clue
to our final destiny, can at least be used in considering our ulti-

mate form of structure. We can conceive of no situation that will remove our desire for rooms, well heated, lighted and with sanitary and hygienic conveniences. It is certain that in the cities at least, the grouping of rooms into a single structure has a fixity from which all conjecture must proceed. The demonstrated advantages of common sources of heat, light and water, the common use of thoroughfares, and the universal access to common media of communication indicate that these will ever be extended, but only to serve the basic requirement of human convenience as it remains intrenched in its sheltered and conveniently equipped rooms.

All of these things point in the same direction so far as construction is concerned—ever larger and more efficient structures, with conveniences that will always continue to develop and refine. Good thoroughfare arrangement, with due regard to ease of swift movement from place to place, goes hand in hand with increasing construction. Like our rooms, our city blocks have not greatly changed from time immemorial. True, avenues have widened and straightened, easier circulation has been forced upon us, sometimes reluctantly, but nothing has arisen greatly to change the average requirement, and the city block may be regarded as about fixed. Certain it is that the metropolitan tendency is toward the construction of buildings occupying whole city blocks; already we have many of this kind throughout the land, and the movement is well established as the next great economic phase of construction. The demonstrated economies and conveniences of this latest development herald the advent of the city of single city-block structures.

This leaves only the moot question of height and height limitation to be considered. Limitation of height of metropolitan structures has never been a more acute question than it is to-day. When the first skyscrapers were built their critics denounced them as structurally unsafe, and dismissed them as

capricious, temporary freaks that would soon fall down and
thus seal their own doom. When this prophecy was unful-
filled, and one skyscraper commenced to shoulder another
along our busy thoroughfares, the hue and cry against them as
destroyers of air and light was raised, and to some purpose.
Before anything could be done about it, however, some of our
most cherished avenues of travel almost overnight became
yawning chasms into which the sunlight never penetrated.
The law, with leaden steps, slowly focussed its attention upon
this condition, and we commenced to get our height limitation and zoning laws. Hardly had this been accomplished
when the problem of traffic congestion became the most acute
aspect of metropolitan existence, and to-day that staggering
perplexity of city life overshadows every other problem in importance. It almost threatens the very existence of the convenient if complex living that our city so ideally serves in all
other respects.

It is futile to point the finger of accusation toward any one
phase of city life and condemn that phase in particular as responsible. The responsibility is itself a great complexity to
which many activities of metropolitan existence contribute.
All that may be said is that over-tall buildings contribute some
indefinite and undefinable share to the problem, and to some
extent height limitation is not only justifiable but necessary.
It is a fair guess that the great metropolitan problem of the future will centre around height limitation considered in the light
of street arrangement and the solution of the traffic problem.
Dreamers have vexed the question by injecting the possibilities
of aviation into it, and already the fantasy of the skyscraper
landing roof is portrayed in our Sunday supplements. The
imagined fulfilment of these dreams contributes nothing to
the solution of the question, for these fantasies simply add another aspect to it.

As an escape, some theorists are actually visioning an abandonment of the great cities that the skyscraper has made, and the construction of new centres, with Utopian arrangements that seem perfectly to meet the requirements of our many methods of swift communication. Perhaps even these wild conjectures may be realized in some now unthinkable way; but if they are, if new and wholly different cities are built, if new and wholly undreamed-of means of transportation and communication are devised, if wholly different building materials are invented, and refinements of convenience developed beyond our wildest conjectures, yet the basic human requirement will be the same. Until human nature, and even human existence itself, is changed, that basic requirement will be shelter, light, heat, sanitation, and swift transportation and communication; and architects, engineers and builders will be in demand to study and solve these problems. Truly, building construction is the most fundamental requirement of human progress.

INDEX

Lightning Source UK Ltd.
Milton Keynes UK
UKHW020645170519
342856UK00005B/516/P

9 781375 924313